T0339907

LINER SHIP FLEET PLANNING

LINER SHIP FLEET PLANNING

Models and Algorithms

TINGSONG WANG
Wuhan University, Wuhan, PR China

SHUAIAN WANG
The Hong Kong Polytechnic University, Kowloon, Hong Kong

QIANG MENG
National University of Singapore, Singapore

Elsevier
Radarweg 29, PO Box 211, 1000 AE Amsterdam, Netherlands
The Boulevard, Langford Lane, Kidlington, Oxford OX5 1GB, United Kingdom
50 Hampshire Street, 5th Floor, Cambridge, MA 02139, United States

Library of Congress Cataloging-in-Publication Data
A catalog record for this book is available from the Library of Congress

British Library Cataloguing-in-Publication Data
A catalogue record for this book is available from the British Library

ISBN: 978-0-12-811502-2

For information on all Elsevier publications visit our website at https://www.elsevier.com/books-and-journals

Publisher: Joe Hayton
Acquisition Editor: Tom Stover
Editorial Project Manager: Anna Valutkevich
Production Project Manager: Priya Kumaraguruparan
Cover Designer: Mark Rogers

Typeset by SPi Global, India

CONTENTS

ABOUT THE AUTHORS

Tingsong Wang is an associate professor in the Department of Economics and Management at Wuhan University (China). His research interests include maritime transportation, network design and optimization, with a focus on modeling, and algorithm design for liner ship fleet planning.

Shuaian Wang is an associate professor in the Department of Logistics and Maritime Studies at Hong Kong Polytechnic University. His research interests include maritime transportation, container shipping and port operations, and transportation network modeling and analysis.

Qiang Meng is a professor in the Department of Civil and Environmental Engineering at National University of Singapore. His research expertise includes transportation network modeling and optimization, shipping network analysis, intermodal freight transportation, and quantitative risk analysis of transport operations.

PREFACE

Compared with the huge number of publications and researches on land transportation, which is the traditional research field of transportation, the researches on maritime transportation attract much less attention. However, maritime transportation is quite important in practice, as maritime transportation fulfills over 70% of global trade. Specifically, liner shipping transports 50% of maritime cargos in monetary terms due to its regular and reliable service and the high value of goods transported. To date, very few academic book chapters on maritime transportation modeling and optimization can be found. Based on our contact with the maritime research community and professionals from shipping companies, there is a need for them to have a book that systematically introduces the new progress in the study of liner ship fleet planning. We aimed to write a book that is suitable for readers with only basic quantitative backgrounds. In this book, we provide readers with systematic approaches to investigate problems and with methodologies for addressing other shipping issues. Hence, the book will be valuable for maritime academics and shipping professionals alike.

In the book, we intend to introduce the latest researches on liner shipping. In detail, the features and content that will be most valuable to reader are as follows:

1. We analyze the problems of liner ship fleet planning and point out that addressing problems should take uncertainty into account because there are always uncertain factors in practice such as uncertain demand. With the consideration of uncertainty, these problems are totally different than those studied by the existing researches, and viable methodologies are required to model these different problems.
2. As for the problems with consideration of uncertainty, we introduce how to use stochastic programming models to formulate them, and we propose the solution algorithms to solve the proposed models.
3. The stochastic programming modeling methods and solution algorithms addressed in this book can not only be used in liner shipping, but also be extended to other research fields that should take uncertainty into account, such as container yard planning with uncertain demand, berth allocation with uncertain vessel arrival times, vehicle deployment problems with uncertain demand, and aircraft assignment problems with uncertain demand.

Our intention was to present the stochastic programming based methodologies and their applications in liner ship planning. We hope that this book will prove useful to students, researchers, and engineers, who encounter work involving uncertainty analyses. We also expect that this book could encourage many to undertake research in this exciting and particularly important field.

Tingsong Wang
Shuaian Wang
Qiang Meng

ACKNOWLEDGMENTS

This book could not have been undertaken without incurring numerous debts. Unfortunately, space limitations and shoddy record keeping do not allow us to name everyone who has contributed to this book. However, we wish to particularly appreciate the following companies for providing us with numerous practical insights: American President Lines (APL), China COSCO Shipping, CMA CGM, Orient Overseas Container Line (OOCL), Pacific International Lines (PIL), and Zhonggu Shipping. We are grateful to Mr. Eddie K H Ng from APL, Ms. Sabrina Sun from NYK Line, and Mr. Goh Hung Song from Centre for Maritime Studies at National University of Singapore for sharing their valuable working experiences in the liner shipping industry with us. Nevertheless, responsibility for the materials contained in the book lies with us.

The first author thanks the Humanities and Social Science Youth Foundation of Ministry of Education of China that supported the development of this book under the research grant No. 16YJC630112. The second author is supported by the research grant 1-ZE5J from the Hong Kong Polytechnic University. The third author is supported by the research project "Liner Shipping Container Slot Booking Patterns and Their Applications to the Shipping Revenue Management" (WBS No. R-302-000-177-720) funded by NOL Fellowship Programme of Singapore. These financial supports are gratefully acknowledged.

Finally, we express our appreciation to Elsevier's acquisitions editor, Tom Stover, for the support and advice he has given throughout the editing and publication process. Our appreciation also extends to editorial project managers, Amy Invernizzi, Anna Valutkevich, and Priya Kumaraguruparan for their kind reminders throughout the writing process.

Tingsong Wang
Shuaian Wang
Qiang Meng

Introduction

CHAPTER ONE

Introduction to Shipping Services

Contents

SEABORNE TRADE

Development of Seaborne Trade

Seaborne trade refers to goods that are transported by ships. It is the main artery, in a sense, of international trade, standing at the apex of the world's economic activity. The increasing globalization and interdependence of various world economies are leading to tremendous positive growth in the seaborne trade industry. According to the maritime transportation report released in 2015 by the United Nations Conference on Trade and Development (UNCTAD) secretariat, the volume of world seaborne shipments in 2014 increased by 3.4% over 2013. This increase is moderate because of the slow-moving recovery of the world economy, which is due to both an uneven economic recovery rate in developed countries and a slowdown in the economic growth rate in developing countries. The world's economic growth still has significant impact on the seaborne trade volumes; even the responsiveness of merchandise trade to the world's GDP growth is moderate in recent years. According to the review of maritime transport produced by UNCTAD, international seaborne trade increased from 5.983

http://dx.doi.org/10.1016/B978-0-12-811502-2.00001-6

Table 1.1 Developments in International Seaborne Trade, Selected Years (millions of tons loaded)

Year	Oil and Gas	Main Bulks[a]	Other Dry Cargo	Total
1970	1440	448	717	2605
1980	1871	608	1225	3704
1990	1755	988	1265	4008
2000	2163	1295	2526	5984
2005	2422	1709	2978	7109
2006	2698	1814	3188	7700
2007	2747	1953	3334	8034
2008	2742	2065	3422	8229
2009	2642	2085	3131	7858
2010	2772	2335	3302	8409
2011	2794	2486	3505	8784
2012	2841	2742	3614	9197
2013	2829	2923	3762	9514
2014	2826	3112	3903	9842

[a]Iron ore, grain, coal, bauxite/alumina, and phosphate rock; the data for 2006 onwards are based on various issues of the Dry Bulk Trade Outlook, produced by Clarksons Research.
Sources: UNCTAD secretariat, based on data supplied by reporting countries and as published on relevant government and port industry websites, as well as by specialist sources.

billion tons in 2000 to 9.842 billion tons in 2014, showing a 3% annual average growth rate during the last 4 years (Review of Maritime Transportation, 2015). The trend in the growth rate of international seaborne trade for selected years is shown in Table 1.1.

Structure of Seaborne Trade

In addition to the analysis of the growth rate of seaborne trade, we can analyze the structure of seaborne trade as well. Dry cargo refers to the cargo that is of solid, dry material; it generally excludes cargo requiring special temperature controls. As can be seen in Table 1.1, the volume of dry cargo has accounted for over two-thirds of the total, which indicates that dry cargo is the main type of seaborne trade. The shipment of dry cargo in 2014 reaches to a total of 7015 millions of tons, which increased 5% compared with the shipment of dry cargo in 2013. At the same time, it still occupies a major proportion in the total goods loaded, which is more than 71% in 2014. Meanwhile, the share of tanker trade, including petroleum products, crude oil, and gas, has slightly declined from 29.8% in 2013 to 28.7% in 2014 (see Tables 1.2 and 1.3).

Table 1.2 World Seaborne Trade 2006–14, by Type of Cargo (Millions of Tons)

	Goods Loaded				Goods Unloaded			
Year	Total	Crude	Petroleum Products and Gas	Dry Dargo	Total	Crude	Petroleum Products and Gas	Dry Cargo
2006	7700.3	1783.4	914.8	5002.1	7878.3	1931.2	893.7	5053.4
2007	8034.1	1813.4	933.5	5287.1	8140.2	1995.7	903.8	5240.8
2008	8229.5	1785.2	957.0	5487.2	8286.3	1942.3	934.9	5409.2
2009	7858.0	1710.5	931.1	5216.4	7832.0	1874.1	921.3	5036.6
2010	8408.9	1787.7	983.8	5637.5	8443.8	1933.2	979.2	5531.4
2011	8784.3	1759.5	1034.2	5990.5	8797.7	1896.5	1037.7	5863.5
2012	9196.7	1785.7	1055.0	6356.0	9188.5	1929.5	1055.1	6203.8
2013	9513.6	1737.9	1090.8	6684.8	9500.1	1882.0	1095.2	6523.0
2014	9841.7	1710.3	1116.1	7015.3	9808.4	1861.5	1122.6	6824.2

Sources: UNCTAD secretariat, based on data supplied by reporting countries, as published on relevant government and port industry websites, and by specialist sources.

Table 1.3 World Seaborne Trade 2006–14, by Type of Cargo (Percentage Share)

	Goods Loaded				Goods Unloaded			
Year	Total	Crude	Petroleum Products and Gas	Dry Cargo	Total	Crude	Petroleum Products and Gas	Dry Cargo
2006	100	23.2	11.9	65.0	100	24.5	11.3	64.1
2007	100	22.6	11.6	65.8	100	24.5	11.1	64.4
2008	100	21.7	11.6	66.7	100	23.4	11.3	65.3
2009	100	21.8	11.8	66.4	100	23.9	11.8	64.3
2010	100	32.1	11.7	67.0	100	22.9	11.6	65.5
2011	100	20.0	11.8	68.2	100	21.3	11.8	66.6
2012	100	19.4	11.5	69.1	100	21.0	11.5	67.5
2013	100	18.3	11.5	70.3	100	19.8	11.5	68.7
2014	100	17.4	11.3	71.3	100	19.0	11.4	69.6

Sources: UNCTAD secretariat, based on data supplied by reporting countries, as published on the relevant government and port industry websites, and by specialist sources.

In 2014, dry cargo accounted for 68% of total world seaborne trade. Among the dry cargo, the dry bulk trade which includes the five major bulk commodities (grain, coal, iron ore, phosphate rock, and bauxite/alumina) and the minor bulk commodities (metals and minerals, agribulks, and manufactures) reached to 4.55 billion tons and accounted for 65 percentage of all dry cargo shipments and "Other dry cargo" (general cargo, break bulk, and containerized) accounted for 35 percentages. Within "other dry cargo,"

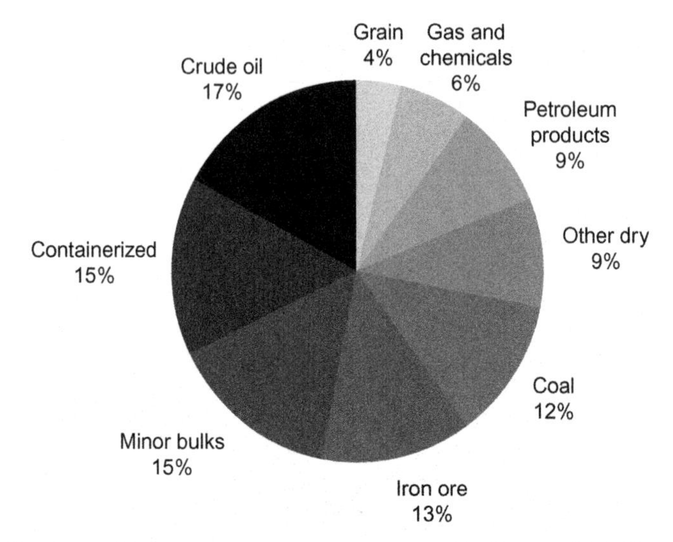

Fig. 1.1 Structure of international seaborne trade, 2014. *(Source: UNCTAD secretariat, based on Clarksons Research, Seaboren Trade Monitor, 2(5), May, 2015.)*

containerized trade reached to 1.63 billion tons and has a proportion of about 66 percentages, and it accounted for 15 percentages of the total seaborne trade. Within tanker trade, crude oil has the largest proportion, which accounted for over 50%. Fig. 1.1 shows the proportion of different components in the structure of world seaborne trade in 2014.

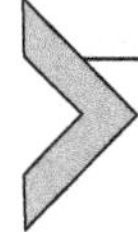

CONTAINER PORTS

Containers

Compared to conventional cargo units, such as boxes, pallets, and cartons, containers have a few significant advantages that enable the rapid growth of containerization. The most important advantage is the reduction of cargo handling time at sea ports. A quay crane can move 20–30 TEUs per hour, and a large containership can accommodate over four quay cranes at the same time. Consequently a ship only spends one day for loading or discharging 2000 TEUs at a port. Before containerization a ship might spend 2 months at a port for cargo handling, which significantly reduced the productivity of ships. For instance, assume that a ship can carry 30,000 tons of cargo and needs 40 days to sail between two ports. If the cargo handling time at a port is 30 days, the ship can transport a total of $365/(30+40) \times 30{,}000 = 156{,}428$ tons of cargo each year. If the cargo handling time

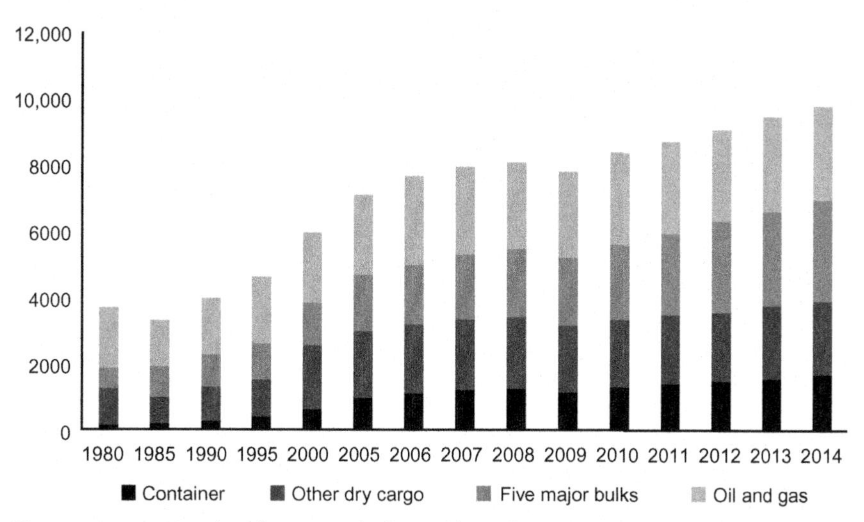

Fig. 1.2 International seaborne trade for a selected year. *(Sources: UNCTAD, Review of Maritime Transport, various issues. For 2006–14, the breakdown by type of cargo is based on Clarksons Research, Shipping Review and Outlook, various issues.)*

at a port is 1 day, the ship can transport a total of $365/(1+40) \times 30{,}000 = 267{,}073$ tons of cargo each year.

The second advantage is intermodalization. Besides dedicated cellular gearless containerships, dedicated trucks and trains are also available for transporting containers in inland areas. Consequently, it is very convenient to move a container of cargo between ships, trains, and trucks.

The last advantage is the reduction of cargo damage and pilferage. Consequently, more and more cargo is being containerized. It can be seen in Fig. 1.2 that containerized trade, when compared with other types of cargo, is growing at the fastest rate in world seaborne trade transportation. A combination of different factors results in the fast growth of containerized trade, such as dedicated purpose-built container vessels, larger vessels capable of achieving increased economies of scale, improved handling efficiency of facilities and equipment in ports, and also the increasing amount and types of raw materials carried in containers (here a container refers to the twenty-foot equivalent unit [TEU]).

Throughput of Container Ports

The number of TEUs loaded and unloaded by a container port is measured as the throughput of the port. The container throughput of the world's top 20 container ports in 2014 and the market share percentage change are listed

Table 1.4 Top 20 Container Ports in the World and Their Throughputs in 2014 (Million TEUs and Percentage Change)

Rank	Port Name	Country	Volume	Percentage Change 2014–2013
1	Shanghai	China	35.29	−3.62
2	Singapore	Singapore	33.87	3.89
3	Shenzhen	China	24.04	3.27
4	Hong Kong	Hong Kong, China	22.20	−0.68
5	Ningbo	China	19.45	12.10
6	Busan	South Korea	18.68	5.61
7	Guangzhou	China	16.61	8.50
8	Qingdao	China	16.58	6.83
9	Dubai	United Arab Emirates	15.20	11.43
10	Tianjin	China	14.06	8.15
11	Rotterdam	Netherlands	12.30	5.83
12	Port Klang	Malaysia	10.95	5.76
13	Kaohsiung	Taiwan Province of China	10.60	6.59
14	Dalian	China	10.13	1.15
15	Hamburg	Germany	9.73	5.09
16	Antwerp	Belgium	8.98	4.66
17	Xiamen	China	8.57	7.04
18	Tanjung Pelepas	Malaysia	8.50	11.43
19	Los Angeles	U.S.A.	8.34	5.99
20	Jakarta	Malaysia	6.05	−1.91

Source: UNCTAD secretariat, based on Dynamar B.V., June 2015, and various other sources.

in Table 1.4 (UNCTAD, 2015). In 2014, the top 20 container ports handled a total of 310.13 million TEUs and occupied the proportion of about 45.7% of the world's container port throughput (see Table 1.4). We can see that within the 20 top container ports, 16 ports are from developing nations in Asia, three ports are from developed countries in Europe, and one port is from a developed country in North America. The rationale behind such a distribution is that countries in Asia, particularly China, are world manufacturing hubs. China has a total of 31,705 berths, which is the most in the world. Consequently the cargo in terms of metric ton volume operated by China is the largest in the world, as is China's cargo in terms of throughput of TEUs. As for the total length of combined navigable rivers, China's waterways reach 126,300 km, longer than those of any other country in the world. In 2014, China handled a total of 12.45 billion tons of cargo, including cargo handled by river ports, which was an increase of 5.8% over the previous year. In 2013, containerized cargo reached

189.85 million TEUs; it increased 6.4% in 2014, reaching 202 million TEUs. China's major ports contributed the most cargo handled by the country, and they operated a total of 2.7 billion tons of cargo in 2014, which increased by 2.2% compared with that of 2013. The increase rate of 2.2% is moderate, mostly because the demand of major commodities is weaker (Yu, 2015).

Operational Performance of Ports

The efficiency and performance of ports and terminals can be used to indicate a country's trade competiveness. If a port is efficient, then its nation can be regarded as competitive; however, if it is congested, then it can be regarded as a barrier to international trade (Van Marle, 2015). Many factors can affect the efficiency and performance of a port/terminal, such as labor relations, number and type of equipment to move cargo, accessibility of port and customs efficiency, and so on. The port operators can improve accordingly the port efficiency and performance based on the indicator of these factors (Cetin, 2015). Table 1.5 shows the top 10 port terminals in 2014 in terms of their berth productivity, with Yokohama ranking as the world's most efficient container port, having reached 180 moves per ship per hour and

Table 1.5 Top Global Terminals' Berth Productivity, 2014 (Container Moves Per Ship, Per Hour on All Vessel Sizes)

Terminal	Port	Country	Berth Productivity
APM Terminals Yokohama	Yokohama	Japan	180
Tianjin Port Pacific International	Tianjin	China	144
DP World-Jebel Ali Terminal	Jebel Ali	United Arab Emirates	138
Qingdao Qianwan	Qingdao	China	136
Tianjin Port Alliance International	Tianjin	China	132
Ningbo Beilun (second)	Ningbo	China	127
Guangzhou South China Oceangate	Nansha	China	122
Busan Newport Co. Ltd	Busan	Republic of Korea	119
Yantain International	Yantain	China	117
Nansha Phase I	Nansha	China	117

Source: JOC Port Productivity Database 2015.

Table 1.6 World's Leading Ports by Productivity, 2014 (Container Moves Per Ship, Per Hour on All Vessel Sizes and Percentage Increase)

Port	Country	2013 Berth Productivity	2014 Berth Productivity	Percentage Increase (%)
Jebel Ali	United Arab Emirates	119	138	16
Tianjin	China	130	125	−4
Qingdao	China	126	125	−1
Nansha	China	104	119	14
Yantian	China	106	117	10
Khor al Fakkan	United Arab Emirates	119	108	−9
Ningbo	China	120	107	−11
Yokohama	Japan	108	105	−3
Busan	Republic of Korea	105	102	−3
Xiamen	China	106	90	−15

Source: UNCTAD secretariat and JOC Port Productivity Database 2015.

having increased productivity by 10% over the previous year. The reason Yokohama terminal ranks at the top of the world in terms of efficiency is that it develops an operation process to make the quay crane operate TEUs synchronally with the operation by yard equipment. Such a synchronized operation between the quay crane and yard equipment can eliminate the waiting time between them.

The top 10 ports with the performance of efficiency and productivities are listed in Table 1.6. As we can see from Table 1.6, in 2014 Yokohoma was the most efficient container port in the world. Besides the reason that the synchronized operation between the quay crane and yard equipment can improve the efficiency and productivity of a port, another possible explanation for making a port more efficient and productive is the intraport competition between different terminal operators on the port. For example, Tianjin port has different terminal operators, including international operators such as APM Terminals, PSA, and DPW, etc., as well as Chinese operations such as China Merchants Holdings International, COSCO Pacific, and CSX World Terminals OOCL, etc. These terminal operators of different nations have to improve their efficiency and productivity in order to survive the competition in transforming the Tianjin port into the most efficient port in China and ranked second place in the world.

MARITIME TRANSPORTATION

Operational Modes

In 2015 the global economy realized that development must change from extensive development to sustainable development. Therefore the year 2015 is regarded as a milestone for sustainable development in history. Maritime transportation is regarded as the backbone of world trade, as it carries around 80% of global trade by volume and over 70% by value (UNCTAD, 2015). Maritime transportation can be divided into three different modes of operation: *industrial*, *tramp,* and *liner shipping.*

In industrial shipping the container owner or the shipper owns the ships and aims to ship all of his/her containers for as low a cost as possible.

In tramp shipping the carrier or tramp shipping company has to carry containers to specified ports in a specific time frame, according to their contracts with shippers. Additional containers (if any are available in the market) are selected depending on the ship's capacity so as to bring in as much revenue as possible.

In liner shipping the carrier releases predetermined maritime routes and schedules to the shippers, and then operates accordingly. In other words, liner shipping provides a fixed liner service at regular intervals between named ports and offers transport to any goods. Time is very important in liner shipping, as the ships have to comply with schedules even when they are operating at low utilization levels.

Thus one can think of industrial shipping as owning a car, tramp shipping as hiring a taxi, and liner shipping as traveling by bus, with definite schedules and a published itinerary. Liner shipping occupies a major position within global transportation. With the continuous advancement of shipbuilding technologies and a greater global container transportation demand, the dominance of liner shipping is expected to continue increasing in strength (UNCTAD, 2015). Many factors have a significant impact on the improvement of the development of the maritime transportation. Within these factors are particular rules adopted by International Maritime Organization (IMO), which have special force to drive maritime transportation to improve its performance in terms of energy, environment. Issues like safety, security, marine pollution, labor conditions, and air pollution all have to be considered by the authorities of maritime transportation in order to sustain development.

Liner Shipping

Liner shipping mainly involves the transportation of containerized cargos. In fact, among all sea cargo, 52% in monetary terms are containerized (UNCTAD, 2015). Containerized cargo, such as electronics, appliances, furniture, garments, auto parts, and toys, generally have much higher unit values than other sea cargo. The total container trade volume in 2015 amounted to 151 million TEUs (UNCTAD, 2015).

Containers are usually transported by liner shipping services with fixed sequences of ports of call at regular service frequencies, which are published on the liner shipping companies' websites in advance to notify shippers. Shippers or freight forwarders can pick up and deliver their cargo at any port covered by the liner services. A single shipper usually has far less than a full shipload of cargo, whereas containerships have to adhere to their published departure dates, even when a full payload is not available.

One can appreciate liner shipping services by likening them to bus services. First, bus services have fixed routes and stops. Liner shipping services also have fixed routes and fixed sequences of port visits. Buses transport many passengers at the same time (in contrast to taxis), and containerships also transport containers from many customers at the same time. At each bus stop, there are passengers alighting and boarding, and at each port there are containers discharged and loaded.

There are also a number of differences between liner shipping services and bus services. First, liner shipping services are usually provided weekly, therefore the arrival day at each port of call is published to facilitate inland transportation. For instance, suppose that ships arrive in Pusan on Thursday. If the cargo from Seoul missed the scheduled ship, it must wait until next Thursday. By contrast, bus service in metropolitan areas has a service frequency of 5–15 min, for example. Therefore it does not matter much to miss a bus. Moreover, buses have many stops and are subject to the uncertain traffic conditions in the city. Hence, the uncertainities of arrival times may be greater than the average waiting time due to the regular service frequency. Consequently, many bus services do not publish the planned arrival time at each bus stop. Suburban bus services are much more infrequent, with a service frequency of 30 min or 1 h, for example. As a result, many suburban bus services have a schedule of arrival times at each bus stop. Second, liner shipping services transport containers while bus services transport customers. Passengers choose their own routes from origin to destination, whereas containers can be transported according to a schedule that is optimal for the transport system. Third, in bus services, the boarding, alighting, and transfer of passengers do not incur any cost. However, in liner shipping, loading, the discharging and transshipment of containers are very expensive, and they are the main revenue for port operators. Fourth, buses usually operate in the daytime, whereas containerships operate 24 h a day and 7 days a week. Fifth, bus driver scheduling is a challenging problem for bus companies, as each bus must have a driver, and drivers have scheduled shifts. By contrast the crew of a containership works on the ship for a long period of time.

The top 20 liner shipping companies in terms of market share, total shipboard capacity deployed (TEUs), number of vessels, and average vessel size on May 1, 2015 are shown in Table 1.5. As can be seen in Table 1.7 the three largest liner shipping companies, APM-Maersk, MSC, and CMA, have a

Table 1.7 Top 20 Liner Shipping Companies, May 1, 2015 (Number of Ships and Total Shipboard Capacity Deployed, Ranked by TEU)

Rank	Operator	Market Share % (TEU)	TEU	Vessels	Average Vessel Size
1	APM-Maersk	13.45	2,526,490	478	5286
2	MSC	13.22	2,486,979	451	5508
3	CMA CGM S.A.	8.00	1,502,417	375	4006
4	Evergreen Line	5.08	954,280	204	4678
5	COSCO Line	4.55	854,171	158	5406
6	CSCL	4.00	751,507	136	5526
7	Hapag-Lloyd	3.90	732,656	145	5053
8	Hanjin Shipping	3.41	640,490	104	6159
9	MOL	3.19	599,772	111	5403
10	APL	2.91	545,850	96	5686
11	OOCL	2.77	520,328	103	5052
12	Hamburg Süd Group	2.66	498,902	104	4797
13	NYK	2.63	494,953	104	4759
14	Yang Ming Corp.	2.60	487,771	103	4736
15	HMM	2.13	399,791	65	6151
16	"K" Line	2.12	397,623	77	5164
17	PIL	1.99	374,849	139	2697
18	UASC	1.98	372,841	53	7035
19	Zim	1.58	296,554	66	4493
20	CSAV	1.26	237,567	40	5939

Note: Includes all container-carrying ships known to be operated by liner shipping companies.
Source: UNCTAD secretariat, based on data provided by Lloyd's List Intelligence.

share of almost 35% of the world's total container-carrying capacity. All of the top three companies are headquartered in Europe (Denmark, Switzerland, and France), while most other carriers among the top 20 are based in Asia.

Chartering ships in maritime transportation is quite common, because owning a ship indicates a huge capital investment. It is noted that ship chartering is adopted by most shipping operators, because it is quite a huge investment to own ships. In fact, some liner companies own less than half of the ships they operate.

REFERENCES

Cetin, C. K. (2015). Port and logistics chains: Changes in organizational effectiveness. In D. W. Song & P. Panayides (Eds.), *Maritime logistics: A guide to contemporary shipping and port management.* (2nd ed.). London: Kogan Page.

UNCTAD (2015). Review of maritime transportation 2015. In: *Paper presented at the United Nations conference on trade and development. New York and Geneva.* http://unctad.org/en/PublicationsLibrary/rmt2015_en.pdf. (Accessed 23 April 2015).

Van Marle, G. (2015). Measuring port performance. *Long Read, 1.* Available at http://theloadstar.co.uk/wpcontent/uploads/The-Loadstar-LongRead-Port-productivity1.pdf (Accessed 22 September 2015).

Yu, A. (2015). Chinese ports handled 202 million TEU in 2014. *Journal of Commerce.* Available at http://www.ihsmaritime360.com/article/17726/chinese-ports-handled-202-million-teu-in-2014 (Accessed 14 September 2015).

CHAPTER TWO

Liner Ship Fleet Planning

Contents

INTRODUCTION

Many researchers have paid attention to the problem of liner ship fleet planning (LSFP) and have studied it for a long time. Their research can be categorized into three groups. The first focuses on an optimal ship fleet design, including determining the numbers and types of ships needed in a fleet over a particular planning horizon, given a set of liner ship routes and the required regular frequency of liner shipping service for each route. Given a fleet of heterogeneous ships and a set of liner ship routes, the second group focuses on optimal fleet deployment, which covers the assignment of ships to each route according to the required regular frequency of service required to satisfy the container shipping requirements. The third group

Liner Ship Fleet Planning
http://dx.doi.org/10.1016/B978-0-12-811502-2.00002-8

focuses on a joint optimal ship fleet design and fleet deployment plan; that is, given a set of liner ship routes, decisions are made regarding the numbers and types of ships and ship assignment to routes in order to satisfy the container shipping requirements. The problems tackled by each of these three groups are referred to through this chapter as the liner ship fleet size and mix problem, the liner ship fleet deployment problem, and the LSFP problem, respectively.

Tackling these problems has become a key task for both the liner operators and researchers. In the past, liner operators relied mainly on their experience and common sense to choose the best plan from a limited set of alternatives. The task is not always difficult, especially when there aren't many options; however, when large fleets are involved, there are numerous options, making it quite difficult to choose the best among them. Empirically based selection strategies are too cumbersome; hence the focus has switched to analysis-based strategies for all three of these problems. Some mathematical programming models and algorithms have been proposed. Most of the related research is surveyed in six review articles: Ronen (1983, 1993); Perakis (2002); Christiansen, Fagerholt, and Ronen (2004); Christiansen, Fagerholt, Nygreen, and Ronen (2013); and Meng, Wang, Andersson, and Thun (2014). As liner shipping is one of three modes of maritime transportation, the following review sections do not strictly focus on LSFP problems; they include the problems for the other two modes as well.

FLEET SIZE AND MIX

Fleet size and mix problems are defined as follows: Given a set of routes, the planner must decide on the exact ship types to include in the fleet, their sizes, and the number of ships of each size. The analytical models built for fleet size and mix problems can be divided into three classes: linear programming models, integer programming models, and dynamic programming models. There are also some simulation models used as decision support systems, in practice. These four types of model are reviewed in the following sections.

Linear Programming Models

Dantzig and Fulkerson (1954) were pioneers in applying the linear programming approach to the fleet size problem. In this article, they aimed at minimizing the number of tankers required to meet a fixed schedule and formulated this problem as a linear programming model solved using the simplex algorithm.

Lane, Heaver, and Uyeno (1987) presented a linear programming model for determining a cost-efficient fleet, meeting the known demand for trade between Australia and the North American West Coast, which incorporates six ports. This problem was dealt with by separating it into three major phases:

Phase I: Voyage Option Enumeration
Phase II: Vessel Scheduling
Phase III: Set Partitioning

Phase I is a combinatorial problem that depends on the number of ports on the trade route. In this phase, all feasible itinerary options are enumerated. A feasible itinerary is defined as including at most one ballast or deadhead leg. Phase II is the key component of the problem, which is to make cost-minimizing trade-off decisions for vessel scheduling at every origin port. A forward-looking heuristic method is used to decide which cargo will be transported by which route, and the algorithm proceeds to determine the cost-minimized (late-loading cost) schedules for port arrivals and departures. Phase III uses the results from Phase II to define the most efficient fleet composition by means of a set partitioning algorithm used to select a subset of the route options, which satisfy all shipping demands at the lowest possible cost.

Integer Programming Models

Fagerholt (1999) proposed a three-phase approach for finding the optimal fleet and coherent routes for that fleet. They studied a homogeneous fleet; that is, only one type of ship is considered in this paper, and a weekly service frequency is required for each shipping route. Phase I is to generate feasible single routes for the ships with the largest capacities. A single route is defined to be a route that is feasible with respect to the vehicle routing problem (VRP) constraints; that is, it originates and terminates at the depot and without visits in between. Phase II is to combine the single routes generated in Phase I into multiple routes. In Phase III, the problem for finding the optimal fleet and coherent routes for that fleet is formulated as a set-partitioning problem, as below:

$$\min \sum_{r \in R} \left(C_r^{TC} + C_r^{OP} \right) x_r \tag{2.1}$$

subject to

$$x_r \in \{0, 1\}, \quad \forall r \in R \tag{2.2}$$

$$\sum_{r \in S^k} x_r \leq N^k, \quad \forall k \in K, \tag{2.3}$$

where R is defined as the set of all routes (both single and multiple) generated in Phases I and II, indexed by r; N is defined as the set of nodes or ports to be serviced by the fleet of ships, indexed by i; C_r^{TC} is the fixed time-charter cost; C_r^{OP} is the operational cost of route r for the lowest-cost ship that has sufficient capacity to perform the given route; x_r is a binary variable which is equal to one if route r is chosen in the optimal solution and zero otherwise; S^k denotes the set of routes for ship type k; and N^k denotes the maximum number of available ships of type k. In order to generate the routes in Phases I and II, the route generation algorithms are written and compiled in Borland Pascal 7.0 in the numerical experiment. The set partitioning model proposed in Phase III is implemented and solved by commercial optimization software of GAMS/CPLEX 5.0 on a PC.

Fagerholt and Lindstad (2000) studied a real problem of determining an efficient policy involving the optimal fleet and corresponding weekly schedules for a supply vessel operation in the Norwegian Sea. The operation involves one onshore service depot located on the northwest coast of Norway and seven offshore installations located in the Norwegian Sea. Six scenarios are developed, in which the opening hours and number of weekly services of the installations are varied, and the best policy is obtained by evaluating the qualitative aspects of the solution for each scenario. The solution algorithm includes two steps for each given scenario. In the first step a number of feasible candidate schedules are generated for each vessel in the pool. The duration of each schedule is also generated; this consists of the sailing times, the loading/discharging and waiting times at the offshore installations, and the turnaround time at the depot. In the second step, the vessels to be used and their weekly schedules are determined by solving an integer programming model. Finally, a scenario is recommended that incurs the least cost for operating the supply vessels.

Sambracos, Paravantis, Tarantilis, and Kiranoudis (2004) considered a problem in which small containers were dispatched by using coastal freight liners in the Aegean Sea. There is only one depot port (Piraeus) from which containers are dispatched to 12 other ports (islands). A homogeneous fleet is used, and demand is fulfilled so as to incur minimum costs, including fuel consumption and port costs. This problem was solved along two dimensions. At the strategic level of planning, the issue is about how to determine the vessel traffic with a given shipment demand in order to minimize the total costs, including the fuel costs and port dues; the issue is formulated as a linear programming model. The planning problem in this issue is defined on a graph G through a set of ports and containers transported from node i to node j, denoted by W_{ij}. Supposing n_{ij} is the number of ships traveling from node i to node j; $n_{\max}$ is the maximum number of ships in each direction for a link; L_{ij} is the length of link ij (in miles); c_F is the cost of fuel consumption per mile; c_{Pj} is the fee for port i per ship; Q is the capacity of a ship, assumed constant for all ship types; D_i is the demand at port i; and S_i is the supply at port i. The problem is formulated as follows:

$$\min \sum_{i,j} \left[n_{ij} L_{ij} c_F + n_{ij} c_{Pj} \right], \tag{2.4}$$

subject to

$$\sum_j W_{ji} - \sum_k W_{ik} + S_i - D_i = 0 \tag{2.5}$$

$$\left(n_{ij} - 1 \right) Q \leq W_{ij} \leq n_{ij} Q \tag{2.6}$$

$$0 \leq n_{ij} \leq n_{\max} \tag{2.7}$$

$$n_{ij},\ W_{ij} \geq 0. \tag{2.8}$$

Subsequently the operational dimension of the problem is analyzed by introducing a VRP formulation corresponding to the periodic needs for transportation using smaller containers, and a list-based threshold acceptance algorithm is employed to solve this problem.

Dynamic Programming Models

Nicholson and Pullen (1971) studied a ship fleet management problem that concerned phasing out a fleet of general cargo ships over a 10-year period with the possibility of premature sales and temporary replacement by charter ships. The objective was to determine a sale and replacement policy that maximized the long-term assets of the company. The method was based on two stages: The first stage determines an order of priority for selling the ships, regardless of the rate at which charter ships are taken on. The second stage uses dynamic programming to determine an optimal level of chartering, given the order of priority for replacement. The first stage essentially reduces the dynamic programming calculation from an N-state variable problem to a one-state variable problem, which is a computationally manageable using a dynamic programming method. The orders of replacement for different ships are different as well, and they have priorities. The order of priority for replacement is evaluated as the following two steps: In the first step the selling price of a ship is used as an assessment to determine the priority order of this ship. With the priority order obtained in the first step, a dynamic programming approach is used in the second step to find the best plan of chartering. The net contribution of a ship to the final assets consists of the invested earnings of that ship up to the year when it was sold, plus the invested net realization from selling it in that year, plus any earnings from a charter ship taken on in lieu of that ship for a limited period. The bigger the net contribution, the higher the ship's order of priority. Let $f_t(j)$ be the maximum cash assets accumulated at the end of year t if j ships are held in year t and an optimal policy has been adopted. Let $g_t(i,j)$ be the increase in cash assets in year t if i ships are held in year $t-1$ and j ships are held in year t. Then the dynamic programming recurrence relations between $f_t(j)$ and $f_{t-1}(i)$ can be set up as follows:

$$f_t(j) = \max(f_{t-1}(i)(1+r) + g_t(i,j)), j \leq i \leq D_{t-1}, \qquad (2.9)$$

where D_{t-1} is the total number of ships required in year $t-1$, and $g_t(i,j)$ consists of the earnings in year t from owned and chartered ships, plus the receipts from sales. The dynamic programming recurrence relations are evaluated for $j = M_t$ to D_t and for $t = 1, \ldots, T$ in turn, setting $f_0(i) = 0$. For each evaluation of the recurrence relation the best value of i is recorded; for example $q_t(j)$. If the largest value of $f_{T+1}(j)$ occurs for $j = x_T$ ships to be held in year T, and in general the number of ships to be held in year t is $x_t = q_{t+1}(x_{t+1})$, then the dynamic programming procedure combined with the order of priority will determine the ships to be held in each year, using the results $x_1, x_2, \ldots, x_T$. Ship numbers N, $N-1, \ldots, x_1+1$ are sold in year 1, numbers x_1, x_1-1, x_2+1 in year 2 and so on and $D_t - x_t$ ships are chartered in year t.

Simulation Models

Stott and Douglas (1981) developed a software system called Marine Operations Planning and Scheduling System (MOPASS) that is used by shipping companies to plan and schedule shipments of bulk goods. This model is a collection of integrated models that provide comparisons of voyage costs for different vessels and trades, a financial evaluation and optimization of vessel-to-trade assignment, and the sequencing and scheduling of individual vessels on predefined routes. The main purpose of MOPASS is to evaluate the most profitable opportunities which may arise for a given controlled fleet of vessels. MOPASS comprises four major subsystems: a linear programming optimization module embedded in one of the subsystems, user-oriented information files, and reports for both management and operating personnel. These various components of MOPASS are accessed, shared, and integrated as needed through a user-oriented executive control program. The model does not deal with the question of overall fleet efficiency, but rather with short-run dynamic operations associated with a given fleet and trade opportunities.

Gallagher and Meyrick (1984) developed a cost-based simulation model designed to analyze the economic characteristics of liner shipping services on a trade route. The model initially defines the components of the shipping system; that is, vessels, ports, trade requirements, and trade routes. Next, cargo assignments are made according to user preference rules and vessel availability. The cargo allocation is then adjusted to obtain feasibility, and finally, the costs of the system are estimated. This simulation model described by Gallagher and Meyrick (1984) is similar to that proposed by Stott and Douglas (1981); it is evaluative as well. Unlike MOPASS, it quantifies system performance with a view to improving the efficiency of the entire shipping system. However, the model does not use a formal optimization model; rather, it focuses on evaluating changes to the existing system.

FLEET DEPLOYMENT

Fleet deployment problems are described as follows: Given a set of ships and a set of routes, the planner must assign the vessels to specific trade routes (i.e., this is a tactical problem). These problems also include the determination of the expected number of layup days (if any) for each ship each year. The analytical approaches to fleet deployment problems can be classified into three types: the linear programming approach, the nonlinear programming approach, and the integer programming model. Again, there are some simulation models that are used for fleet deployment in practice. These four types of model are introduced in the following sections.

Linear Programming Models

Laderman, Gleiberman, and Egan (1966) considered a problem of how to assign ships in order to carry the commodities required by customers, and they formulated this problem as a linear programming model. Given a set of ports and vessels, this paper aimed to minimize the total operating time or maximize the total unused time such that the fleet carried out the customers' shipment requirements:

$$\max \sum_{k} z_k \tag{2.10}$$

subject to

$$\sum_{i,j} T_{ij}^k X_{ij}^k + \sum_{i,j} t_{ij}^k x_{ij}^k + z_k = T_k \text{ (for all } k) \tag{2.11}$$

$$\sum_{k} V_{ij}^k X_{ij}^k = A_{ij} \left(\text{for all } i, j, \text{ such that } A_{ij} > 0\right) \tag{2.12}$$

$$\sum_{j} X_{ij}^k = \sum_{j} x_{ij}^k \text{ (for all } i, k) \tag{2.13}$$

$$\sum_{i} X_{ij}^k = \sum_{i} x_{ij}^k \text{ (for all } j, k) \tag{2.14}$$

$$X_{ij}^k, x_{ij}^k \geq 0 \text{ (for all } i, j, k) \tag{2.15}$$

$$z_k \geq 0 \text{ (for all } k), \tag{2.16}$$

where A_{ij} is the amount to be shipped from origin i to destination j (tons); V_{ij}^k is the tonnage capacity of vessel k when going from origin i to destination j; T_{ij}^k is the total time required for vessel k to load at i, go from i to j, and unload at j; t_{ij}^k is the time required for an empty vessel k to go from j to i, T_k is the time available for vessel k during the shipping season; X_{ij}^k is the number of loaded trips to be made by vessel k from origin i to destination j, x_{ij}^k is the number of empty trips to be made by vessel k from destination j to origin i; and z_k is the amount of "slack" or unused time for vessel k. In the paper the decision variables X_{ij}^k and x_{ij}^k are relaxed into continuous variables.

Bradley, Hax, and Magnanti (1977) studied a mission and composition problem, and they formulated this problem as a linear programming model. The objective was to determine the number of ships of different types, as well as the voyages that satisfied the annual shipping requirements ("mission") on a defined set of possible routes, and at a minimum present value cost. However, a number of simplifying assumptions had to be made in order to facilitate the formulation of this model as a linear programming problem. The restrictions of the linear framework thus limit its accuracy in modeling specific shipping services.

Nonlinear Programming Models

Benford (1981) developed a nonlinear programming model for selecting the most profitable fleet deployment strategy while satisfying customer demands by means of a trial and error method. The objective of the procedure was to select the mix of available ships and sea speeds that would perform the required service at maximum profitability to the owner. This paper considered a route with only two ports in which goods are transported directly between these two ports. The paper assumed that there were more than enough ships to meet the customers' demands, and that there were no appreciable costs or benefits involved in taking excess ships out of service. It first estimated the economic characteristics of each ship when operated at a range of reduced speeds, which involved the annual transport capacity (tons), annual operational cost, unit transportation cost and corresponding speeds. Then, it searched for the minimum operating cost by means of trial and error.

Perakis (1985) used the Lagrangian method to resolve this problem and obtained a solution that was better in terms of solution quality. Based on Benford (1981), the annual capacity of a ship was assumed to be a linear function, and the associated operating cost per ton was a quadratic function with respect to speed. This gave the annual capacity of a ship of type i operating at speed x_i to be

$$\alpha_i x_i + \beta_i, \tag{2.17}$$

and the operating cost per ton was given by

$$\gamma_i x_i^2 + \delta_i x_i + \varepsilon_i. \tag{2.18}$$

Hence the operating costs for each ship per year can be computed by using the below equation:

$$a_i x_i^3 + b_i x_i^2 + c_i x_i + d_i = (\alpha_i x_i + \beta_i)(\gamma_i x_i^2 + \delta_i x_i + \varepsilon_i) \tag{2.19}$$

The objective function is

$$\min \sum_{i=1}^{N} n(i)(a_i x_i^3 + b_i x_i^2 + c_i x_i) + \sum_{i=1}^{N} n(i) d_i \tag{2.20}$$

subject to

$$\sum_{i=1}^{N} n(i)(\alpha_i x_i) = C_0 - \sum_{i=1}^{N} n(i)\beta_i \tag{2.21}$$

The problem can be equivalently stated by using Lagrange multipliers, as follows:

$$\min L = \min \sum_{i=1}^{N} n(i)\left(a_i x_i^3 + b_i x_i^2 + (c_i + \lambda \alpha_i)x_i\right) + \sum_{i=1}^{N} n(i) d_i - \lambda\left(C_0 - \sum_{i=1}^{N} n(i)\beta_i\right) \tag{2.22}$$

The solution of Eq. (2.22) can be obtained by setting

$$\frac{\partial L}{\partial x_i} = 0 \tag{2.23}$$

$$\frac{\partial L}{\partial \lambda} = 0 \tag{2.24}$$

The paper then supposed that there are N groups of $n(i)$ identical ships and that C_0 (tons) is the required annual carrying capacity between two given ports. From Eq. (2.24), we have

$$x_i = \frac{\sqrt{b_i^2 - 3a_i(c_i + \lambda\alpha_i)} - b_i}{3a_i}, \quad i = 1, \ldots, N. \tag{2.25}$$

Substituting Eq. (2.25) into Eq. (2.21), we get

$$\sum_{i=1}^{N} \frac{n(i)\alpha_i}{3a_i}\left(\sqrt{b_i^2 - 3a_i(c_i + \lambda\alpha_i)} - b_i\right) = C_0 - \sum_{i=1}^{N} n(i)\beta_i. \tag{2.26}$$

Eq. (2.26) can be numerically solved by secant method. The optimal value of λ obtained from Eq. (2.26) is then substituted into Eq. (2.22) to give us the optimal speeds $x_1, \ldots, x_N$.

Perakis and Papadakis (1987a, 1987b) developed a new nonlinear programming model for the same problem as was considered in Benford (1981) and Perakis (1985). Perakis and Papadakis (1987a) divided the speeds of ships into two classes: ballast speeds for the ship when it does not carry cargo and full load speeds when it carries cargo at its largest capacity. The objective was to determine each vessel's full load and ballast speeds such that the total fleet operating cost was minimized and all contracted cargo was transported. Given a fleet of Z ships, each with a given full load cargo-carrying capacity, and each had known operating cost characteristics as functions of vessel speed. For each individual vessel in the fleet, the total operating costs per ton and total tons carried per year over a specific trade route was expressed as two functions with respect to full load and ballast speeds, denoted by $F_i(X_i, Y_i)$ and $G_i(X_i, Y_i)$. Then the total operating cost of a vessel per year was given by

$$C_i(X_i, Y_i) = F_i(X_i, Y_i)G_i(X_i, Y_i). \tag{2.27}$$

The optimization problem was to minimize the annual total operating cost of the fleet on the specified route, as follows:

$$\min \sum_{i=1}^{Z} C_i(X_i, Y_i) \tag{2.28}$$

subject to the following constraints:

$$X_{i\min} \leq X_i \leq X_{i\max}, \quad i = 1, \ldots, Z \tag{2.29}$$

$$Y_{i\min} \leq Y_i \leq Y_{i\max}, \quad i = 1, \ldots, Z \tag{2.30}$$

$$\sum_{i=1}^{Z} G_i(X_i, Y_i) = C_{av} \tag{2.31}$$

where X_i and Yi respectively denote the full load and ballast speed of ship I; $X_{i\max}$ and $X_{i\min}$ denote the upper and lower bounds of X_i, respectively ($Y_{i\max}$ and $Y_{i\min}$ are defined similarly); and C_{av} is the cargo available for transporting by the fleet.

The authors employed the Nelder and Mead Simplex Search Technique and the External Penalty Technique to solve their model. However, the number of round trips obtained using the optimal solution is not an integer. Thus in order to find the optimal solution with an integer number of round trips, a sequential optimization approach was used by Perakis and Papadakis (1987b). These were taken to be the integral part (or the integral part plus one) of the real numbers of round trips obtained. In this paper, one or more costs were also assumed to be random variables with known probability density functions. Those costs were the fuel price, the constant costs (which is the sum of the annual staffing, administrative, maintenance, supply, and equipment costs) for each ship, as well as the port and route charges for each ship. Analytical expressions for the basic probabilistic quantities (i.e., the probability density function) and the mean and variance of the total operating cost were presented in the paper. The expected value of the total operating cost is then computed; the objective of the model is to minimize it as follows:

$$\min \bar{C} = A \cdot \bar{C}_f + B \cdot \bar{C}_{pm} + \sum_{k=1}^{K} (D_k + E_k \cdot \bar{C}_{km}) + F, \tag{2.32}$$

where $\bar{C}_f$, $\bar{C}_{pm}$ and $\bar{C}_{km}$ denote the expected values of the fuel price, the constant costs, and the port and route charges for each ship.

In the above articles (Benford, 1981; Perakis, 1985; Perakis & Papadakis, 1987a, 1987b), the authors considered a fleet deployment problem with one origin and one destination; that is, two specific ports. Papadakis and Perakis (1989) extended this to consider a fleet deployment problem with multiple origins and destinations, studying the problem of minimizing the cost of operating a fleet of ships that has to carry a specific amount of cargo from a set of loading ports (origins) to a set of unloading ports (destinations) in a given time period. The paper formulated the operating cost as a nonlinear function, with respect to the full-load and ballast speeds of the ships in the same way as Perakis and Papadakis (1987a, 1987b) did. The objective function is given by

$$\min \sum_{i,j,k} \left(V_{i,j,k} N_{i,j,k} + U_{i,j,k} M_{i,j,k} \right) + \sum_{k} L_k Z_k, \tag{2.33}$$

subject to

$$\sum_{i,j} \left(\left(\frac{d_{i,j}}{24 X_{i,j,k}} + t_{i,j,k} \right) N_{i,j,k} + \left(\frac{d_{i,j}}{24 Y_{i,j,k}} + t'_{i,j,k} \right) M_{i,j,k} \right) + Z_k = T_k, \quad k = 1, \ldots, K \tag{2.34}$$

$$\sum_{i,k} W_k N_{i,j,k} = B_j, \quad j = 1, \ldots, J \tag{2.35}$$

$$\sum_{j,k} W_k N_{i,j,k} = Q_i, \quad i = 1, \ldots, I \tag{2.36}$$

$$\sum_{i} N_{i,j,k} = \sum_{i} M_{i,j,k}, \quad k = 1, \ldots, K; j = 1, \ldots, J \tag{2.37}$$

$$\sum_{j} N_{i,j,k} = \sum_{j} M_{i,j,k}, \quad k = 1, \ldots, K; \; i = 1, \ldots, I \tag{2.38}$$

$$X_k^{\min} \leq X_{i,j,k} \leq X_k^{\max} \tag{2.39}$$

$$Y_k^{\min} \leq Y_{i,j,k} \leq Y_k^{\max}, \tag{2.40}$$

where $M_{i,j,k}$, $N_{i,j,k}$, $X_{i,j,k}$, $Y_{i,j,k}$ and Z_k are the decision variables, $M_{i,j,k}$ denotes the number of ballast trips for vessel k from port j to port i, while $N_{i,j,k}$ denotes the number of full load trips, $X_{i,j,k}$ and $Y_{i,j,k}$ respectively denote the full-load and ballast speeds, Z_k denotes the idle time for vessel k, $U_{i,j,k}$ is the total operating cost of vessel k traveling from port i to port j in ballast conditions, while $V_{i,j,k}$ is the equivalent under full-load conditions, L_k is the daily lay-up cost, $d_{i,j}$ is the distance between port i and port j, $t_{i,j,k}$ is the time required for vessel k to unballast and load at i plus any time required to travel from port i to port j, $t'_{i,j,k}$ is similar to $t_{i,j,k}$ for unloading at port j plus a ballast trip from port j to port i, T_k is the time available for vessel k during the shipping season, W_k denotes the cargo capacity of vessel k, B_j is the amount of cargo to be delivered to the destination port j, and Q_i denotes the available amount of cargo at source port i. The authors analyzed the properties of their model and found that $Y_{i,j,k}$ could be expressed as a function with respect to $X_{i,j,k}$ and that $X_{i,j,k}$ is the solution to an equation. In other words, the decision variables $X_{i,j,k}$ and $Y_{i,j,k}$ can be eliminated from their model. Finally, they applied a projected, augmented Lagrangian algorithm to find the optimal solution.

Integer Programming Models

Cho and Perakis (1996) considered the liner fleet deployment problem in order to determine the number of ships in the fleet, the types of ship in the fleet and the ship-to-route assignments in the fleet. Given a set of ships the paper aimed to assign each ship to a mix of routes among a finite set of candidate routes so as to minimize the total cost or maximize the total profit. Two programming models were formulated: a linear programming model and a mixed integer programming model. An augmented flow-route incidence matrix was introduced to facilitate the expression of the models. The linear programming model is given by

$$\max \sum_{r=1}^{R} \sum_{k \in K_r} \pi_{rk} x_{rk}, \tag{2.41}$$

subject to

$$\sum_{r=1}^{R} \sum_{k \in K_r} a_{ij,rk} x_{rk} \geq m_{ij}, \quad \forall (i,j) \tag{2.42}$$

$$\sum_{r \in R_k} t_{rk} x_{rk} \leq t_k, \quad k = 1, \ldots, K, \tag{2.43}$$

where x_{rk} is the decision variables (fractions, not integers); π_{rk} denotes the expected profit from a round trip on route r by ship k; $a_{ij,rk}$ is a component of the augmented flow-route incidence matrix; m_{ij} is the minimum required number of trips from port i to port j; t_{rk} is the total travel time for ship k on route r per round trip; t_k is the maximum time ship k is available during the planning horizon; R_k is the set of routes r to which ship k can be assigned; and K_r is a set of available ships that can be assigned to route r. The problem can also be represented in matrix form as follows:

$$\max \pi x \tag{2.44}$$

subject to

$$\bar{A}x \geq m \tag{2.45}$$

$$Tx \leq t. \tag{2.46}$$

Mourão, Pato, and Paixão (2001) presented an application of an integer programming model that could be used to support the decision-making process for assigning ships with hub and spoke constraints, solving the model by means of the MS Excel solver function. In this paper, three levels of ports are identified: The main port is part of the medium-sized transport network that feeds mainline ocean trading and is the principle cargo origin and destination. The hub port represents the consolidation port which links the medium-sized network with the smaller transport network and embeds the terminal ports. Finally, the spoke port is the terminal port, where cargo is delivered to the end consumer.

The ships are classified into two types: mainline and feeder ships. The mainline ships move between the main ports and the hub ports, while the feeder ships link the hubs to each set of spoke ports. Two scenarios are proposed: Scenario A consists of scheduling the main and the feeder ships as if a coordinated voyage situation is anticipated, and it assumes that a fixed number of voyages are performed each year by each ship, whether main or feeder vessels. Scenario B is constructed exclusively to perform a sensitivity analysis of the solution obtained for Scenario A. Hence Scenario B sets out to determine the optimum number of voyages each ship should undertake annually, in accordance with each roster. Scenario A and Scenario B are formulated as an integer programming model, respectively. Finally, MS Excel's solver is employed to solve the two models.

Simulation Models

Some of the simulation models described previously in "Simulation Models" section are also applied to the fleet deployment problem, such as MOPASS (Stott & Douglas, 1981), and the cost-based simulation model (Gallagher & Meyrick, 1984). As they were described earlier, they are omitted from this section. The following paragraph describes a new simulator, developed for fleet deployment problems.

Xie (1997) proposed the Fleet Planning System (FPS), a new simulator that is an optimization-based decision support system for a fleet of heterogeneous vessels aimed at optimizing their deployment and development planning. FPS takes the characteristics of each type of vessel to be known parameters, such as their size, transportation capacity, and costs incurred on each liner trade route. The number of vessels of each type assigned to each route are the decision variables. The minimum cost of shipping the specified and required amount of cargo is the objective, and linear programming techniques are the main method used to optimize the assignment strategy for each vessel, as well as the development planning for the fleet. FPS is coded in the FORTUNE Language and consists of two main programming modules: RDATA and LP. RDATA is designed to read in the initially known data and turn these data into parameters in the linear programming model. LP firstly transforms the linear programming model into a standard linear programming model, then checks the validity of the coefficients of the model and the solvability of the problem. Finally, it optimizes the calculation by means of the simplex algorithm and prints out the results.

FLEET PLANNING

Fleet size and mix problems are strategic-level problems, while fleet deployment problems are tactical-level problems. Agarwal and Ergun (2008) pointed out that the decisions made at one planning level affected the decision-making at the other. At the strategic level, the decisions are to set the general policies and guidelines, which are regarded as regulars that have to be obeyed and followed by the decisions to be made at the tactical level and operational level. In the opposite way the data and information on the shipping costs and revenues obtained by shippers at the lower levels including tactical and operational levels, would give useful and valuable information to decision makers at higher levels. Therefore fleet size, mix problems, andfleet deployment problems are combined by some researchers, who assume that the planner not only decides the fleet size and mix, but also the fleet deployment. These joint problems are referred to as fleet planning problems in this thesis for convenience. Three types of modeling methods have been used in the fleet planning problems and three types of model have been proposed for these problems correspondingly: linear programming, integer programming, and dynamic programming. They are reviewed in the following sections.

Linear Programming Models

Everett, Hax, Lewinson, and Nudds (1972) applied a linear programming approach in order to optimize a fleet of large tankers and bulkers. They proposed the following model to minimize the life-cycle cost of the fleet:

$$\min \sum\nolimits_s n_s I_s + \sum\nolimits_s \left(\sum\nolimits_r C_{sr} x_{sr} + n_s a_s \right) \sum_{t=1}^{T} \left(\frac{1+\beta}{1+\alpha} \right)^t \quad (2.47)$$

subject to

$$\sum\nolimits_s \sum\nolimits_r V_{srk} x_{sr} = d_k \quad (2.48)$$

$$\sum\nolimits_r t_{sr} x_{sr} - 345 n_s \leq 0, \quad (2.49)$$

where n_s and x_{sr} are decision variables denoting the number of ships of type s and the number of voyages per annum assigned to ship s along route r, respectively. I_s is the capital cost for ship type s and C_{sr} is the variable operating cost incurred by ship type s along route r, while a_s is the annual fixed operating cost of ship type s, α is the annual discount rate, and β is the annual inflation rate. T is the length of the planning horizon (years), V_{srk} is the maximum amount of commodity k which can be carried by ship type s along route r, d_k is the total annual tonnage of commodity k specified in the mission for the pertinent pair of ports, and t_{sr} is the time taken to make a round trip by ship type s along route r. All ships are assumed to be available 345 days per year. In order to solve the model, a linear programming system called "Ophelie" is employed to solve the model. The compulsory inputs for the system during the solving progress contain the parameters, which are created by a matrix generator. The system can automatically compute the basic data to yield the final output. However, the solutions are fractions rather than integers.

Perakis and Jarammillo (1991) proposed a linear programming model for an optimal fleet size, mix, and deployment with a detailed cost estimation for liner ships. First of all, they described the costs spent on a round trip, involving port charges, canal fees, fuel costs, maintenance costs, insurance costs, administrative costs, crew costs, and other miscellaneous costs, and then formulated the cost per voyage as a function of the cruising speed of the container ships. The shipping cost per voyage can be expressed by the following equation:

$$C_{kr}(s_{kr}) = \overline{\lambda}_{kr} s_{kr}^2 + \widetilde{\lambda}_{kr} / s_{kr} + \hat{\lambda}_{kr}, \quad (2.50)$$

where $\bar{\lambda}_{kr}$, $\tilde{\lambda}_{kr}$ and $\hat{\lambda}_{kr}$ are parameters. Eq. (2.50) is a convex function, which implies that the optimal cruising speeds of container ships can be obtained so as to minimize shipping costs. Finally, a linear programming model is proposed and formulated as follows:

$$\min \sum_{k=1}^{K} \sum_{r=1}^{R} C_{kr} X_{kr} + \sum_{k=1}^{K} e_k Y_k, \tag{2.51}$$

subject to

$$\sum_{r=1}^{R} t_{kr} X_{kr} + Y_k = 365 N_k^{\max}, \quad \text{for all } k \tag{2.52}$$

$$\sum_{k=1}^{K} X_{kr} \geq \frac{365}{F_r}, \quad \text{for all } r \tag{2.53}$$

$$Y_k \geq (365 - T_k) N_k^{\max} \tag{2.54}$$

$$X_{kr}, Y_k \geq 0 \tag{2.55}$$

where X_{kr} and Y_k are decision variables, which denote the number of annual voyages and layup days of ship k on route r, respectively; e_k is the total daily lay-up cost for ship k; t_{kr} is the voyage time of ship k on route r; F_r is the frequency of service on route r; T_k is the available shipping days for ship k per year; $N_k^{\max}$ is the number of ships of type k available; and K and R are the number of ships and routes considered in the paper. The implementation and results shown in Jaramillo and Perakis (1991) were obtained using the LINDO solver.

Integer Programming Models

Cho and Perakis (1996) proposed a mixed-integer programming model for a long-term ship fleet planning problem. They assumed that a shipping company has to make capital investment decisions over the planning horizon. To meet the expected increasing future cargo demand the shipping company may consider various options for expanding fleet capacity, such as building or purchasing new ships or chartering in existing ships. The objective is to minimize the total cost incurred from operations while meeting the cargo demands over the planning horizon. The total cost included in the objective function is taken to be the sum of the operating cost, the lay-up (idle) cost, and the (fixed) capital cost incurred over the planning horizon. Let K^0 be the subset of ships that the shipping company considers adding to the existing fleet. The resulting objective function is as follows:

$$\min \sum_{r=1}^{R} \sum_{k \subset K_r} c_{rk} x_{rk} + \sum_{k=1}^{K} h_k y_k + \sum_{k \in K^0} f_k z_k, \tag{2.56}$$

subject to

$$\sum_{r \in R_k} t_{rk} x_{rk} + y_k = t_k, \quad \forall k \notin K^0 \tag{2.57}$$

$$\sum_{r \in R_k} t_{rk} x_{rk} + y_k - t_k z_k = t_k, \quad \forall k \in K^0 \tag{2.58}$$

$$\sum_{r=1}^{R} \sum_{k \in K_r} a_{ij,rk} x_{rk} \geq m_{ij}, \quad \forall (i,j) \tag{2.59}$$

$$x_{rk} \geq 0, y_k \geq 0, \quad z_k \in \{0, 1\}, \tag{2.60}$$

where x_{rk} is the set of decision variables (fractions, not integers); y_k is another decision variable, denoting the lay-up time of ship k; and the variable z_k is a binary variable denoting whether ship k will be added to the fleet ($z_k = 1$) ($z_k = 0$). $a_{ij,rk}$ is a component of the augmented flow-route incidence matrix, c_{rk} is the expected operating cost of ship k on route r per round trip, f_k denotes the fixed capital cost involved in adding ship k to the existing fleet, h_k is the lay-up cost of a ship per unit of time, t_{rk} is the total travel time for ship k on route r per round trip, t_k is the maximum time ship k is available during the planning horizon, m_{ij} is the minimum required number of trips from port i to port j, R_k is the set of routes r to which ship k can be assigned, K_r is a set of available ships that can be assigned to route r, and K^0 is the subset of ships which the company considers adding to the fleet.

In the numerical example of Perakis and Jarammillo (1991), the number of ships assigned to each route are not all integers; some are real numbers, which is inconsistent with practice. A rounding procedure was therefore required to transform the number of ships allocated to each route into an integer. The rounding led to some variations in the targeted service frequencies, resulting in suboptimal results. In order to eliminate any rounding errors in the linear programming formulation, Powell and Perakis (1997) reformulated this problem and proposed an integer programming model that they solved by means of an OSL solver. The integer programming optimization model for this problem is formulated as follows:

$$\min \sum_{k=1}^{K} \sum_{r=1}^{R} C'_{kr} N_{kr} + \sum_{k=1}^{K} Y_k e_k, \tag{2.61}$$

subject to

$$\sum_{r=1}^{R} N_{kr} \leq N_k^{\max} \quad \text{for each of type } k \tag{2.62}$$

$$\sum_{k=1}^{K} t'_{kr} N_{kr} \geq M_r \quad \text{for all } r \tag{2.63}$$

$$Y_k = 365 N_k^{\max} - T_k \sum_{r=1}^{R} N_{kr} \tag{2.64}$$

$$N_{kr} \geq 0, \tag{2.65}$$

where N_{kr} and Y_k are the decision variables denoting the number of ships of type k operating on route r and the number of lay-up days per year for a ship of type k, respectively; C'_{kr} is the operating cost of a type k ship on route r; e_k is the total lay-up costs per day for a type k ship; M_r is the number of voyages required per year on route r; t'_{kr} is the yearly voyages made by a ship of type k on route r; and T_k is the duration of the shipping season for a ship of type k. Finally, this linear programming model can be solved by optimization software solvers like CPLEX and LINDO. Here it is solved by LINDO.

Gelareh and Meng (2010) looked at model development for a LSFP problem. First, a mixed-integer nonlinear programming model was presented. Then the proposed nonlinear model was transformed by means of a linearization technique, and a mixed-integer programming model was obtained that can be solved efficiently using a standard mixed-integer programming solver such as CPLEX. The mixed-integer programming model determines the optimal route service frequency pattern and takes into account the constraints that the shipping service time should not exceed a given value.

Dynamic Programming Models

Xie, Wang, and Chen (2000) studied the fleet planning problem for a long-term planning horizon with an approach of dynamic programming. Due to the strategy involved in fleet planning, a horizon of several years can naturally be deconstructed into a series of consecutive decisions made at the beginning of each year. The problem is broken down into two optimal subproblems: one is to apply the annual optimal fleet deployment plan if fleet and transport demand is fixed, and the other is to apply the optimal strategy for fleet development in consecutive years. Therefore the first subproblem can be formulated as a linear integer programming model which seeks the optimal fleet deployment for a short-term planning horizon (one year) and the second subproblem can be formulated as a dynamic programming model, seeking the best liner fleet size and mix over a long-term planning horizon. In the dynamic programming model, one year is taken as one stage, and the quantitative composition of a fleet in terms of ships of various types is taken to be the state of the fleet. The optimization of fleet deployment for the first subproblem can be written as follows:

$$\min Z_{ti} = \begin{cases} \sum_{j=1}^{K} \left(\sum_{h=1}^{G} X_{jhti} R_{jht} + O_{jti} F_{jt} \right) & \text{if } \bar{X}_j \in \bar{\Omega} \\ \infty & \text{if } \bar{X}_t \in \overline{\Psi} \end{cases}, \tag{2.66}$$

where $\bar{X}_t$ is the decision vector, denoting the deployment scheme of ships in year t. Each element X_{jhti} denotes the number of ships of type j distributed on route h in the ith state in year t, and $\bar{\Omega}$ is a set of $\bar{X}_t$, which meets the following two groups of constraints:

$$\sum_{j=1}^{K} X_{jhti} V_{jht} = W_{ht}, \quad h = 1, \ldots, G \tag{2.67}$$

$$\sum_{h=1}^{G} X_{jhti} + O_{jti} = U_{j,t-1} - WT_{jt} + C_{jti}, \quad j = 1, \ldots, K, \tag{2.68}$$

where $\overline{\Psi}$ is the complementary set of $\bar{\Omega}$, C_{jti} is the number of ships of type j added to the fleet in the ith state at the beginning of year t, F_{jt} is the annual lay-up costs for a ship of type j in year t, O_{jti} is the number of laid-up ships of type j in the ith state in year t, R_{jht} is the annual running costs of a ship of type j on route h in year t, $U_{j,t-1}$ is the number of ships of type j before the start of year t, V_{jht} is the annual transportation capacity of a ship type j on route h in year t, W_{ht} is the annual transportation demand on route h in year t, and WT_{jt} is the number of ships of type j that are scrapped or out of commission in year t. The accumulated sum of the costs of running the fleet in the ith state from year t to year N, ZP_{ti}; that is, the recursive formulation, is given by:

$$ZP_{ti} = ZP_{t+1,i} + \frac{Z_{ti}}{(1+\alpha)^t} + \frac{1}{(1+\alpha)^t} \sum_{j=1}^{K} C_{jti} S_{jt} - \frac{\beta}{(1+\alpha)^N} L(N-t) \tag{2.69}$$

where $L(N-t)$ denotes the physical residual value at the end of the planning horizon of the new ships that were added into the fleet in year t, S_{jt} is the market price for a ship of type j in year t, α is the discount rate, and β is the weight coefficient. Finally, the optimal strategy over the whole planning horizon can be obtained, which is the solution of the following optimization model:

$$\min_{i=1,\ldots,M} ZP_{0i}, \tag{2.70}$$

where M represents the number of various combinations of ships that can be added to the fleet at the beginning of year 0. In order to solve the model formulated for the problem, a heuristic algorithm is developed.

RESEARCH LIMITATIONS AND GAPS

It can be seen from the literature review that there are some limitations and gaps in the existing studies. This section highlights these and shows how they provoke the need for further investigation. The limitations of past studies fall under the following three types:

Firstly and most importantly, all of the previous works reviewed above assume an environment in which the container shipment demand between port pairs is known beforehand. The container shipment demand between port pairs is a deterministic input, and it can be obtained by using traditional forecasting methods. However, they can never be forecasted with complete confidence because they are affected by some unpredictable and uncontrollable factors; therefore, it is more reasonable to regard the demand as uncertain.

Secondly, the parameters of annual operating cost and transportation capacity of each ship on each route are assumed to be constants in Cho and Perakis (1996) and Xie et al. (2000). Such an assumption is unreasonable because it is inconsistent with reality. In fact, these parameters should be voyage-dependent. If the numbers of ships sailing on a route in one year are different, then the operating costs of these ships on the route for the year is certainly different. A greater number of ship sailing indicates that there will be higher operating costs, and it also implies that more containers and cargo would be transported by the ship, namely the capacity of a ship that sails more trips is larger than a ship that sails less trips.

Thirdly, the methodology proposed by Cho and Perakis (1996) for a multiperiod ship fleet planning problem is unreasonable. In their methodology, once the decisions about fleet design and deployment are made at the beginning of the planning horizon, these decisions are assumed to be fixed and static over the whole multiperiod planning horizon. However, the strategy should be dynamic, especially the decisions in regard to the number of ships, types of ships, and ship-to-route assignments should be adjustable based on the current situation of the fleet on each period.

Based on the limitations observed in previous studies, it is realistic and necessary to take the uncertainty of container shipment demand into account in LSFP problems. As a result, the LSFP problem could become a new and interesting research topic and provide a fresh angle on the classical LSFP problem which is studied under a deterministic environment. The models proposed for classical LSFP problems in previous studies cannot be used directly here. Therefore the first purpose of this book is to propose new models for LSFP problems with container shipment demand uncertainty, then to propose effective solution algorithms for solving the new models.

In addition, this book revises the unreasonable assumptions described above in order to consider a more realistic LSFP problem than has been studied previously in the literature. Moreover, it provides an applicable and feasible way for a liner container shipping company to carry out its LSFP in practice.

SUMMARY

This chapter has presented a critical literature review focusing on three problems: fleet size and mix problems, fleet deployment problems, and fleet planning problems. Through this review, several potential problems and gaps have been identified. Finally, the chapter described the research purpose of this book.

REFERENCES

Agarwal, R., & Ergun, O. (2008). Ship schedule and network design for cargo routing in liner shipping. *Transportation Science, 42*(2), 175–196.

Benford, H. (1981). A simple approach to fleet deployment. *Maritime Policy and Management, 8*(4), 223–228.

Bradley, S. P., Hax, A. C., & Magnanti, T. L. (1977). *Applied mathematical programming.* Reading, MA: Addison-Wesley. Chapter 7.

Cho, S. C., & Perakis, A. N. (1996). Optimal liner fleet routing strategies. *Maritime Policy and Management, 23*(3), 249–259.

Christiansen, M., Fagerholt, K., Nygreen, B., & Ronen, D. (2013). Ship routing and scheduling in the new millennium. *European Journal of Operational Research, 228*(3), 467–478.

Christiansen, M., Fagerholt, K., & Ronen, D. (2004). Ship routing and scheduling: Status and perspectives. *Transportation Science, 38*(1), 1–18.
Dantzig, G. B., & Fulkerson, D. R. (1954). Minimizing the number of tankers to meet a fixed schedule. *Naval Research Logistics Quarterly, 1*, 217–222.
Everett, J. L., Hax, A. C., Lewinson, V. A., & Nudds, D. (1972). *Optimization of a fleet of large tankers and bulkers: A linear programming approach*: (pp. 430–438). Marine Technology.
Fagerholt, K. (1999). Optimal fleet design in a ship routing problem. *International Transactions in Operational Research, 6*(5), 453–464.
Fagerholt, K., & Lindstad, H. (2000). Optimal policies for maintaining a supply service in the Norwegian Sea. *Omega, 28*(3), 269–275.
Gallagher, F. D., & Meyrick, S. J. (1984). *ASEAN-Australia liner shipping: A cost-based simulation analysis*. Kuala Lumpur/Canberra: ASEAN-Australia Joint Research Project. ASEA-Australia economic papers no. 12.
Gelareh, S., & Meng, Q. (2010). A novel modeling approach for the fleet deployment problem within a short-term planning horizon. *Transportation Research Part E, 46*(1), 76–89.
Jaramillo, D. I., & Perakis, A. N. (1991). Fleet deployment optimization for liner shipping Part 2. Implementation and results. *Maritime Policy and Management, 18*(3), 235–262.
Laderman, J., Gleiberman, L., & Egan, J. F. (1966). Vessel allocation by linear programming. *Naval Research Logistics Quarterly*, 315–320.
Lane, D. E., Heaver, T. D., & Uyeno, D. (1987). Planning and scheduling for efficiency in liner shipping. *Maritime Policy and Management, 12*(3), 109–125.
Meng, Q., Wang, S., Andersson, H., & Thun, K. (2014). Containership routing and scheduling in liner shipping: Overview and future research directions. *Transportation Science, 48* (2), 265–280.
Mourão, M. C., Pato, M. V., & Paixão, A. C. (2001). Ship assignment with hub and spoke constraints. *Maritime Policy and Management, 29*(2), 135–150.
Nicholson, T. A. J., & Pullen, R. D. (1971). Dynamic programming applied to ship fleet management. *Operational Research Quarterly, 22*(3), 211–220.
Papadakis, N. A., & Perakis, A. N. (1989). A nonlinear approach to the multiorigin, multidestination fleet deployment problem. *Naval Research Logistics, 36*, 515–528.
Perakis, A. N. (1985). A second look at fleet deployment. *Maritime Policy and Management, 12* (3), 209–214.
Perakis, A. N. (2002). Fleet operation optimization and fleet deployment. In C. Th. Grammernos (Ed.), *The handbook of maritime economics and business*. London: Lloyds of London Publications. 580–597.
Perakis, A. N., & Jarammillo, D. I. (1991). Fleet deployment optimization for liner shipping Part 1. Background, problem formulation and solution approaches. *Maritime Policy and Management, 18*(3), 183–200.
Perakis, A. N., & Papadakis, N. (1987a). Fleet deployment optimization models Part 1. *Maritime Policy and Management, 14*(2), 127–144.
Perakis, A. N., & Papadakis, N. (1987b). Fleet deployment optimization models Part 2. *Maritime Policy and Management, 14*(2), 145–155.
Powell, B. J., & Perakis, A. N. (1997). Fleet deployment optimization for liner shipping: An integer programming model. *Maritime Policy and Management, 24*(2), 183–192.
Ronen, D. (1983). Cargo ships routing and scheduling: Survey of models and problems. *European Journal of Operational Research, 12*(2), 119–126.
Ronen, D. (1993). Ship scheduling: The last decade. *European Journal of Operational Research, 71*(3), 325–333.
Sambracos, E., Paravantis, J. A., Tarantilis, C. D., & Kiranoudis, C. T. (2004). Dispatching of small containers via coastal freight liners: The case of the Aegean Sea. *European Journal of Operational Research, 152*(2), 365–381.

Stott, K. L., & Douglas, B. W. (1981). A model-based decision support system for planning and scheduling ocean-borne transportation. *Interfaces*, *11*(4), 1–10.

Xie, X. L. (1997). *Fleet management and deployment*. Beijing: Renmin Jiao Tong Press, ISBN 7-114-03490-3 [in Chinese].

Xie, X. L., Wang, T. F., & Chen, D. S. (2000). A dynamic model and algorithm for fleet planning. *Maritime Policy and Management*, *27*(1), 53–63.

PART TWO

Mathematical Modeling

CHAPTER THREE

Introduction to Stochastic Programming

Contents

MOTIVATION

Uncertainty is often faced by decision makers with tasks such as vehicle scheduling, fleet deployment, financial planning, etc. They are just few examples of areas in which uncertainty has to be taken into consideration. If uncertainty is ignored, it can lead to inferior or downright wrong decisions. The issue of the optimization problem with uncertainty attracts the attention of researchers, motivating them to propose new methodologies and formulations to deal with the issues. As the methodologies and formulations for the optimization problems with certainty cannot be applied any longer, the area of stochastic programming was created. Today, there are different ways to deal with the uncertainty, and various approaches to optimization problems under uncertain environment were developed. This book mainly introduces two models of stochastic programming to deal with the optimization problem with uncertainty: one is chance constrained model in which the uncertainty is formulated as a constraint with probabilistic form; and another one is two-stage model in which decision variables are divided into two stages. Readers who are interested in stochastic programming can refer to the Handbook by Ruszczyński and Shapiro (2002) for more information. In order to motivate the main concepts of stochastic programming, we start with the following classic example: the newsvendor problem.

Liner Ship Fleet Planning
http://dx.doi.org/10.1016/B978-0-12-811502-2.00003-X

EXAMPLES

Newsvendor Problem

Suppose a newsvendor purchases a quantity x of newspapers from a distributor at the beginning of a day at the cost of c per unit. A newspaper is sold at the price of s per unit, and unsold newspapers can be returned to the vendor at the price of r per unit (without generality, we can assume $0 \leq r < c < s$). The demand, that is, the quantity of newspapers needed by customers is denoted by D. The profit earned by the newspaper is then a function of x and D. If it is denoted by $F(x, D)$, it is obviously a piecewise function:

$$F(x, D) = \begin{cases} (s-c)x, & \text{if } x \leq D \\ sD + r(x-D) - cx, & \text{if } x > D \end{cases}. \tag{3.1}$$

Eq. (3.1) can be rewritten as follows:

$$F(x, D) = \begin{cases} (s-c)x, & \text{if } x \leq D \\ (r-c)x + (s-r)D, & \text{if } x > D \end{cases}. \tag{3.2}$$

The profit function can be illustrated as in Fig. 3.1. As it shows that the profit function is with a positive slope $s - c$ for $x < D$ and a negative slope

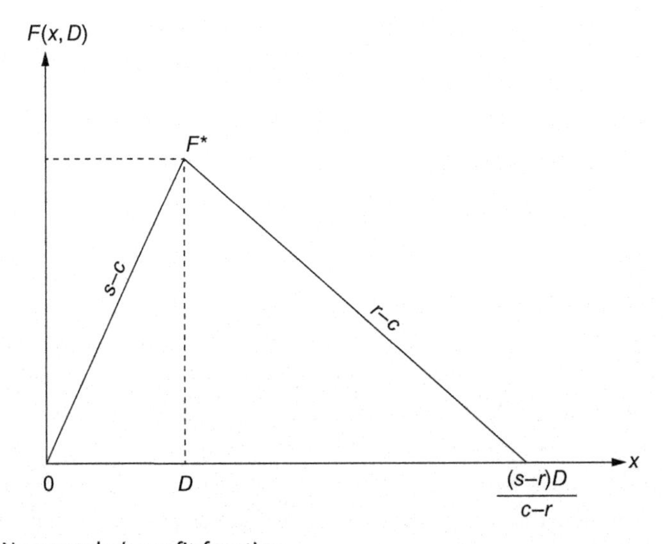

Fig. 3.1 Newsvendor's profit function.

$r - c$ for $x > D$, therefore if the demand D is known, then the best decision for the newsvendor is to choose the order quantity $x^* = D$ in order to get the maximal profit F^*.

However, in reality the demand of newspapers (i.e., D) is unknown at the time when the newsvendor makes the order decision. According to the historical information, D can be viewed as a random variable with a known, or at least a well estimated probability distribution measured by the corresponding cumulative distribution function (CDF) $G(w) := \mathbb{P}(D \leq w)$. It is noted that $G(w) = 0$ for $w < 0$ because the demand cannot be negative. As D is reviewed as a random variable, then $F(x, D)$ is random as well. Therefore it makes sense for the newsvendor to change the objective in order to maximize the expected profit $\mathbb{E}[F(x, D)]$. This leads to the following stochastic programming problem:

$$\underset{x \geq 0}{\text{Max}} \{f(x) := \mathbb{E}[F(x, D)]\}. \tag{3.3}$$

Such models like in Eq. (3.3) are called *expected value models*. If here we treat x as a continuous rather than discrete variable, we have

$$\begin{aligned} f(x) &= \int_0^{\infty} F(x, w) dG(w) \\ f(x) &= \int_0^{x} [(r-c)x + (s-r)w] dG(w) + \int_x^{\infty} (s-r) x dG(w) \\ f(x) &= (s-c)x - (s-r) \int_0^{x} G(w) dw. \end{aligned} \tag{3.4}$$

As the function $f(\cdot)$ is concave and continuous, and it follows that $f(\cdot)$ is differentiable at x if and only if $G(\cdot)$ is continuous at x, then the optimal solution x^* to the problem (3.3) is that $f'(x^*) = 0$. Note that because $r < c < s$, it follows that $0 < (s-c)/(s-r) < 1$. Consequently an optimal solution of Eq. (3.3) is given by

$$x^* = G^{-1}\left(\frac{s-c}{s-r}\right). \tag{3.5}$$

This holds even if $G(\cdot)$ is discontinuous at x^*. It is noted that if $G(0)$ is positive and $G(0) \geq (s-c)/(s-r)$, then the optimal solution $x^* = 0$ because $G(0)$ is equal to the probability that the demand D is zero.

It can be seen that the above approach obviously depends on the knowledge of the probability function of the demand D. But unfortunately, it is quite difficult for us in practice to know exactly the corresponding CDF $G(\cdot)$.

In the present case the optimal solution (3.5) is given in a closed form, and therefore its dependence on $G(\cdot)$ can be easily evaluated.

Now we use another optimization approach for the decision-making problem under uncertainty. In this approach, the random variable D is replaced by its mean $\mu = \mathbb{E}[D]$; then we can obtain a deterministic optimization problem as follows:

$$\underset{x \geq 0}{\text{Max}}\, F(x, u). \tag{3.6}$$

The solution to Eq. (3.6) is denoted by $\bar{x}$, which is often called the expected value solution. In the present example the optimal solution of this deterministic optimization problem (3.6) is $\bar{x} = \mu$, which is quite different from the solution x^* given in Eq. (3.5). The extreme value of CDF of $G(\cdot)$ affects the mean value, while it has no impact on the quantiles, namely the quantiles are much more stable than the mean value to the variations of CDF of $G(\cdot)$ than the corresponding mean value. Therefore, the optimal solution x^* of the stochastic optimization problem is more robust with respect to variations of the probability distributions than an optimal solution $\bar{x}$ of the corresponding deterministic optimization problem.

As for any x, $F(x, D)$ is concave in D; according to Jensen's inequality, we have

$$F(x, \mu) \geq \mathbb{E}[F(x, D)]. \tag{3.7}$$

Hence

$$\underset{x \geq 0}{\text{Max}}\, F(x, \mu) \geq \underset{x \geq 0}{\text{Max}}\, \mathbb{E}[F(x, D)]. \tag{3.8}$$

The above formulates the newsvendor problem with uncertain demand of newspaper as an expected value model to maximize the expected profit for the newsvendor.

Now we formulate this problem from another point view: The newsvendor is also interested in making at least a specified amount of money, denoted by b, on a particular day. As the demand is uncertain, then it would be reasonable for the newsvendor to consider this problem as to purchase the minimum number of newspapers meanwhile the probability of profit at least b is not less than a given value, say $1 - \alpha$, where $\alpha \in (0, 1)$. Such a problem can be formulated in the following form:

$$\text{Min}\, x \tag{3.9}$$

$$\text{s.t.}\, \mathbb{P}\{F(x, D) \geq b\} \geq 1 - \alpha. \tag{3.10}$$

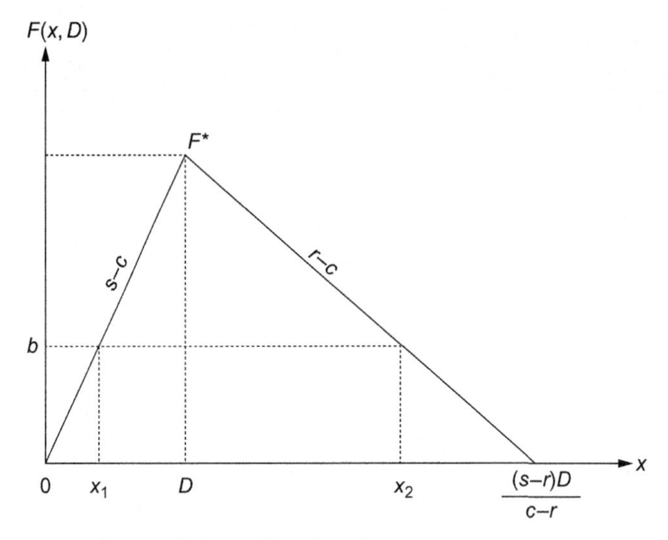

Fig. 3.2 Newsvendor's profit is not less than b.

The constraints, such as in Eq. (3.10), are called *probabilistic* or *chance constraints*. The profit, which is not less than b, can be illustrated in Fig. 3.2. It is clear that the following inequalities should be satisfied:

$$(s-c)x \geq b \tag{3.11}$$

$$(r-c)x + (s-r)D \geq b. \tag{3.12}$$

According to inequality (3.12), we have

$$D \geq \frac{b + (c-r)x}{(s-r)}. \tag{3.13}$$

As for a fixed x satisfying (3.11) the profit $F(x, D)$ is a nondecreasing function of the demand D; therefore according to Eqs. (3.12) and (3.13), we have

$$\mathbb{P}\{F(x, D) \geq b\} = \mathbb{P}\left\{D \geq \frac{b + (c-r)x}{(s-r)}\right\}. \tag{3.14}$$

Let $G^{-1}(\alpha)$ denote the $(1-\alpha)$ – quantile of the distribution of D; then we can get

$$\frac{b + (c-r)x}{(s-r)} \leq G^{-1}(\alpha). \tag{3.15}$$

It is clear that the solution can exist if the constraints (3.11) and (3.15) are consistent; that is, if

$$b \le (s-r)G^{-1}(\alpha). \tag{3.16}$$

Therefore we obtain that problem (3.9) and (3.10) is feasible if Eq. (3.16) holds, in which case it has the optimal solution

$$\hat{x} = \frac{b}{s-c}. \tag{3.17}$$

Investment Problem

Suppose that there are n investment projects, with random returns $R_1, \ldots, R_n$ for each unit capital investment in the future. Now we have an amount of W_0 initial capital, and our aim is to invest x_i to these projects $(i=1, \ldots, n)$ in such a way that the returns on our investments are maximized. As the returns of these projects are random, following Example 1, here it is reasonable for us to maximize the expected returns under the condition that the chance of losing no more than a fixed amount of $b>0$ is at least $1-\alpha$, where $\alpha \in (0, 1)$. Therefore it can be formulated as follows:

$$\begin{aligned} &\text{Max} \sum_{i=1}^{n} \mu_i x_i \\ &\text{s.t. } \mathbb{P}\left\{ \sum_{i=1}^{n} R_i x_i \ge -b \right\} \ge 1-\alpha, \end{aligned} \tag{3.18}$$

where $\mu_i = \mathbb{E}[R_i]$. Note that for the sake of simplicity, we do not impose here the constraint $\sum_{i=1}^{n} x_i = W_0$.

The return R_i is assumed to have a normal distribution for project $i(i=1, \ldots, n)$, and the summation of these n projects is denoted by $G(x, R) = \sum_{i=1}^{n} R_i x_i$. According to the additive property of the normal distributions, $G(x, R)$ is a joint normal distribution with the expected value $\mu^T x$ and variance $x^T \sum x$ (here, μ and x are vectors corresponding to the expected value μ_i and capital investment x_i, and $\sum$ is the covariance matrix). Consequently, $G(x, R)$ can be transformed into a standard normal distribution; that is:

$$\frac{\sum_{i=1}^{n} R_i x_i - \mu^T x}{\sqrt{x^T \sum x}} \sim N(0, 1). \tag{3.19}$$

The chance constraint in Eq. (3.18) is therefore equivalent to the following inequality:

$$\frac{b + \mu^T x}{\sqrt{x^T \sum x}} \geq z_\alpha, \tag{3.20}$$

where z_α is the $(1-\alpha)-$quantile of the standard normal distribution. Therefore Eq. (3.18) can be rewritten as follows:

$$\begin{aligned} &\text{Max} \sum_{i=1}^{n} \mu_i x_i \\ &\text{s.t. } z_\alpha \sqrt{x^T \sum x} - \mu^T x \leq b \end{aligned} \tag{3.21}$$

The problem (3.21) can be solved by methods of convex programming to find the optimal solution. It should be pointed out that the problem (3.18) can be equivalently transformed into Eq. (3.21) because it is based on an assumption that the uncertain parameters in this problem follows normal distributions. Otherwise, it is quite difficult for us to handle such constraints; therefore it is hard to solve the problem in Eq. (3.18), either theoretically or computationally.

This example is simplified because it is implied that the return period is a single one. Now let us assume that the return period is T years, and let $R_i(t)$ be the random investment return for the unit capital investment to the project $i(i = 1, \ldots, n)$ in years $t = 1, \ldots, T$. Note that we just invest at the beginning of the planning horizon, not year after year. Our current aim is to maximize the total investments after T years, and we hope that the returns of our investments will never drop by more than b from the initial amount invested with a probability at least $1-\alpha$.

Let $S_i(t) := \prod_{\tau=1}^{t} (1 + R_i(\tau)) - 1$ be the compounded return of investment to project i up to year t, and denote $\mu_i := \mathbb{E}[S_i(T)]$. Our problem is then as follows:

$$\begin{aligned} &\text{Max} \sum_{i=1}^{n} \mu_i x_i \\ &\text{s.t. } \mathbb{P}\left\{ \sum_{i=1}^{n} S_i(t) x_i \geq -b, t = 1, \ldots, T \right\} \geq 1 - \alpha. \end{aligned} \tag{3.22}$$

The constraint in Eq. (3.22) is called a *joint chance* or *probabilistic constraint*. It requires that for each year, the probability of losing no more than b is at least $1-\alpha$. If we define $\bar{G}(x,R):=\min_{1\le t\le T}\sum_{i=1}^{n} S_i(t)x_i$, then we can rewrite (3.22) as follows:

$$\begin{aligned}&\text{Max}\sum_{i=1}^{n}\mu_i x_i\\&\text{s.t. } \mathbb{P}\{\bar{G}(x,R)\ge -b\}\ge 1-\alpha.\end{aligned}\tag{3.23}$$

The problem with joint chance constraints is much more difficult to handle. As for the solving method, we will talk about it in detail in Chapter 4.

SUMMARY

This chapter presents two classical examples to introduce the mathematical modeling of stochastic programming problems. We have different approaches from different views to formulate an optimization problem with uncertain parameters. The models of stochastic programming problems are difficult to handle either theoretically or computationally.

REFERENCE

Ruszczyński, A., & Shapiro, A. (2002). *Handbooks in operational research and management science.* (Vol. 10). North Holland: Elsevier Science Press.

CHAPTER FOUR

Chance-Constrained Programming

Contents

INTRODUCTION

Chance-constrained problems were first studied and proposed by Charnes in 1958 (Charnes, Cooper, & Symmonds, 1958). Since then, chance-constrained programming has been applied in many areas, such as water management (Dupacová, Gaivoronski, Kos, & Szántai, 1991) and optimization of chemical processes (Henrion et al., 2001; Henrion & Möller, 2003). If readers are interested in the theoretical background of chance-constrained programming, we refer them to Prékopa (1995).

In Chapter 3, we introduced the classical example of the newsvendor problem and formulated it as a chance-constrained programming model. The programming can be expressed as the general form:

$$\begin{aligned} &\min_{x\in X} f(x) \\ &\text{s.t. } \operatorname{prob}\{G(x,\xi)\le 0\}\ge 1-\alpha. \end{aligned} \tag{4.1}$$

Here $X\subset\mathbb{R}^n$, ξ is a random vector with probability distribution P supported on a set $\Xi\subset\mathbb{R}^d$, $\alpha\in(0,1)$, $f:\mathbb{R}^n\to\mathbb{R}$ is a real valued function and $G:\mathbb{R}^n\times\Xi\to\mathbb{R}^m$.

Liner Ship Fleet Planning
http://dx.doi.org/10.1016/B978-0-12-811502-2.00004-1

It has been over 50 years since chance-constrained programming was proposed, yet little progress in the methods to solve it was made until recently. Two reasons result in that the chance-constrained programming model is extremely difficult to solve. One is that for a given $x \in X$, the quantity prob$\{G(x, \xi) \leq 0\}$ is hard to compute, as it requires a multidimensional integration. Consequently, we may only check the feasibility of a given point $x \in X$ by using the Monte-Carlo simulation. Another is that the feasible set of problem (4.1) can be nonconvex, even if the set X is convex and the function $G(x, \xi)$ is convex in x. Therefore we can divide the solving methodologies into two somewhat different directions: one is to discretize the probability distribution P and then to solve the obtained combinatorial problems (e.g., Dentcheva, Prékopa, & Ruszczyński, 2000; Luedtke & Ahmed, 2008); another approach is to employ convex approximations of chance constraints (e.g., Nemiroski & Shapiro, 2006). The solving methodologies will be introduced in detail in Part III of this book. The following will introduce some selected applications of chance-constrained programming.

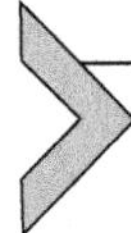

SELECTED APPLICATIONS

Water Resources

Let us consider a water resource problem such as the one studied by Prékopa and Szántai (1978a): There are only two possible reservoir sites in which capacities of each reservoir is denoted by x_1, x_2, and the two reservoirs serve to protect a downstream area from flood that may happen once in a year (see Fig. 4.1). If the water amounts to be retained by the two reservoirs are denoted by ξ_1, ξ_2, then the following equations are hold:

$$\begin{aligned} x_1 + x_2 &\geq \xi_1 + \xi_2 \\ x_2 &\geq \xi_2 \end{aligned} \tag{4.2}$$

Let $c(x_1, x_2)$ denote the cost function of building the two reservoirs. Our aim is to figure out how to build the two reservoirs at the least cost, but they must be effective enough to protect their downstream area from flood, namely satisfying constraints (4.2). Let V_1 and V_2 be upper bounds determined by the local geographic situation of the two reservoir sites. If ξ_1 and ξ_2 are deterministic variables, then we can formulate the water resources problem as follows, called Model I:

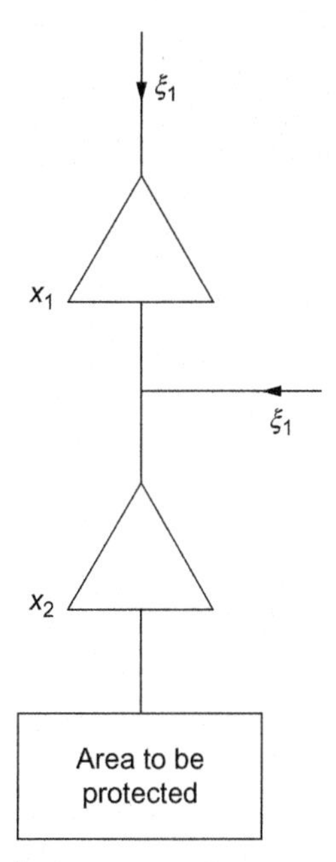

Fig. 4.1 Geographical layout of water reservoirs.

$$
\begin{aligned}
&\min c(x_1, x_2) \\
\text{s.t. } & x_1 + x_2 \geq \xi_1 + \xi_2 \\
& x_2 \geq \xi_2 \\
& 0 \leq x_1 \leq V_1 \\
& 0 \leq x_2 \leq V_2.
\end{aligned}
\tag{4.3}
$$

However, ξ_1 and ξ_2 are obviously random variables because the water to be retained by the two reservoirs are affected by the water from the upstream area including the rain, which are always changeable every year. And consequently, the inequalities $x_1 + x_2 \geq \xi_1 + \xi_2$, $x_2 \geq \xi_2$ may not always hold. As ξ_1 and ξ_2 are random variables, the fulfillment of these inequalities can be guaranteed only on a probability level p, say 95%. Therefore the formulation of the water resource problem with random variables can be developed as follows, called Model II:

$$
\begin{aligned}
&\min c(x_1, x_2) \\
\text{s.t. } & \text{Prob}\begin{pmatrix} x_1 + x_2 \geq \xi_1 + \xi_2 \\ x_2 \geq \xi_2 \end{pmatrix} \geq p \\
& 0 \leq x_1 \leq V_1, 0 \leq x_2 \leq V_2.
\end{aligned}
\tag{4.4}
$$

In order to solve this problem (4.4), Prékopa and Szántai (1978b) assumed that ξ_1 and ξ_2 are following normal and gamma distributions. The chance-constrained programming model was applied for problem of the water level regulation of Lake Balaton in Hungary, and a Gaussian process was used to describe the inflow process. The results indicate that the water lever regulation reliability can be improved from the former 80%–97.5%.

Vehicle Routing Problem

Let us consider the graph of a vehicle routing problem proposed by Dentcheva et al. (2000) and shown in Fig. 4.2, in which there are a total of five nodes: A, B, C, D, and E, representing the cargo demand generation nodes. Each arc in the Fig. 4.2 has two directions. The cargo demands of each node pairs are given in Table 4.1.

There are a total of 19 routes consisting of the arcs shown in Fig. 4.2. Each route is represented by a number from 1 to 19; the route-arc incidence matrix is shown in Table 4.2.

For example, Route 18 contains five arcs: AC, BA, CB, CD, and DC. Therefore Route 18 has the form A→C→D→C→B→A. In addition, the unit cost for a vehicle associated with each route is given by

$$c = (10\ 15\ 18\ 15\ 32\ 32\ 57\ 57\ 60\ 60\ 63\ 63\ 61\ 61\ 75\ 75\ 62\ 62\ 44)$$

Let us take Route 18 as an example again. If a vehicle is assigned to Route 18, then the cost for this vehicle to deliver cargoes along Route 18 from A→C→D→C→B→A is 62. If we define x_i $(i = 1, \ldots, 19)$ as the decision variable representing the number of vehicles assigned on route

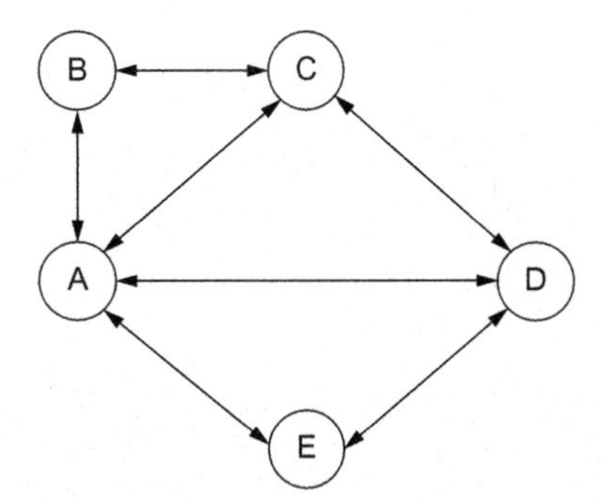

Fig. 4.2 An example of VRP with five sites.

Table 4.1 Expected Values of Traffic Demand of Each Node Pair

Arc	Expected Demand
(A, B)	2
(A, C)	3
(A, D)	2
(A, E)	2
(B, A)	1
(B, C)	1
(C, A)	2
(C, B)	1
(C, D)	4
(D, A)	2
(D, C)	4
(D, E)	3
(E, A)	2
(E, D)	3

Table 4.2 Route-arc Incidence Matrix

Arc	1	2	3	4	5	6	7	8	9	10	11	12	13	14	15	16	17	18	19
AB	1				1						1				1		1		
AC		1				1	1						1					1	
AD			1					1	1			1							
AE				1						1				1		1			1
BA	1					1						1				1		1	
BC					1						1				1		1		
CA		1			1			1						1			1		
CB						1						1				1		1	
CD							1				1		1		1		1	1	
DA			1				1			1	1								
DC								1				1		1		1	1	1	
DE									1				1		1				1
EA				1					1				1		1				1
ED										1				1		1			1

i $(i = 1, \ldots, 19)$, then the total cost of vehicles assigned on all routes can be computed as follows:

$$\begin{aligned}\text{Total cost} = {} & 10x_1 + 15x_2 + 18x_3 + 15x_4 + 32x_5 \\ & + 32x_6 + 57x_7 + 57x_8 + 60x_9 + 60x_{10} \\ & + 63x_{11} + 63x_{12} + 61x_{13} + 61x_{14} + 75x_{15} \\ & + 75x_{16} + 62x_{17} + 62x_{18} + 44x_{19}.\end{aligned} \tag{4.5}$$

In this vehicle routing problem the vehicles are homogenous, namely all vehicles have the same capacity, so we can use $V_{capacity}$ to denote the capacity of vehicle. Let us analyze the capacity on each arc: For example, we use arc AB. We can find that there are a total of five routes containing arc AB according to Table 4.2: Route 1, Route 5, Route 11, Route 15, and Route 17. Therefore the capacity of vehicles on arc AB should include the vehicles assigned to those five routes. If $x_1,\ x_5,\ x_{11},\ x_{15},\ x_{17}$ denote the number of vehicles assigned on the five routes, then the capacity of these vehicles on arc AB can be expressed as follows:

$$\left(x_1 + x_5 + x_{11} + x_{15} + x_{17}\right) \times V_{capacity}. \tag{4.6}$$

Obviously the capacity of vehicles on arc AB shown in Eq. (4.6) should be larger, or at least not less than, the demand on arc AB; namely, we have:

Arc AB:

$$\left(x_1 + x_5 + x_{11} + x_{15} + x_{17}\right) \times V_{capacity} \geq 2. \tag{4.7}$$

For other arcs, we can list the similar constraints as follows:

Arc AC:

$$\left(x_2 + x_6 + x_7 + x_{13} + x_{18}\right) \times V_{capacity} \geq 3 \tag{4.8}$$

Arc AD:

$$(x_3 + x_8 + x_9 + x_{12}) \times V_{capacity} \geq 2 \tag{4.9}$$

Arc AE:

$$(x_4 + x_{10} + x_{14} + x_{16} + x_{19}) \times V_{capacity} \geq 2 \tag{4.10}$$

Arc BA:

$$(x_1 + x_6 + x_{12} + x_{16} + x_{18}) \times V_{capacity} \geq 1 \tag{4.11}$$

Arc BC:

$$(x_5 + x_{11} + x_{15} + x_{17}) \times V_{capacity} \geq 1 \tag{4.12}$$

Arc CA:

$$(x_2 + x_5 + x_8 + x_{14} + x_{17}) \times V_{capacity} \geq 2 \tag{4.13}$$

Arc CB:

$$(x_6 + x_{12} + x_{16} + x_{18}) \times V_{capacity} \geq 1 \tag{4.14}$$

Arc CD:

$$(x_7 + x_{11} + x_{13} + x_{15} + x_{17} + x_{18}) \times V_{\text{capacity}} \geq 4 \tag{4.15}$$

Arc DA:

$$(x_3 + x_7 + x_{10} + x_{11}) \times V_{\text{capacity}} \geq 2 \tag{4.16}$$

Arc DC:

$$(x_8 + x_{12} + x_{14} + x_{16} + x_{17} + x_{18}) \times V_{\text{capacity}} \geq 4 \tag{4.17}$$

Arc DE:

$$(x_9 + x_{13} + x_{15} + x_{19}) \times V_{\text{capacity}} \geq 3 \tag{4.18}$$

Arc EA:

$$(x_4 + x_9 + x_{13} + x_{15} + x_{19}) \times V_{\text{capacity}} \geq 2 \tag{4.19}$$

Arc ED:

$$(x_{10} + x_{14} + x_{16} + x_{19}) \times V_{\text{capacity}} \geq 3. \tag{4.20}$$

The objective of the vehicle routing problem is to minimize the total cost while satisfying the demand on all arcs. Consequently, we have the following mathematical model:

$$\begin{aligned} \min : \text{Total cost} = {} & 10x_1 + 15x_2 + 18x_3 + 15x_4 + 32x_5 \\ & + 32x_6 + 57x_7 + 57x_8 + 60x_9 + 60x_{10} \\ & + 63x_{11} + 63x_{12} + 61x_{13} + 61x_{14} + 75x_{15} \\ & + 75x_{16} + 62x_{17} + 62x_{18} + 44x_{19}, \end{aligned} \tag{4.21}$$

s.t. constraints (4.7)–(4.20).

Now we consider another version of this vehicle routing problem with a stochastic demand. If the demand of arcs are not deterministic but random, then they are random variables following some distribution, such as Poisson distribution, and we assume the values given in Table 4.1 are the expected value of demand on each arc. As the cargo demands are random while the capacity of vehicles is deterministic, there would be a case that the capacity of the vehicles on arcs cannot satisfy the cargo demand. Such a case may happen with a probability, such as 10%. In other words the capacity of vehicles on arcs can satisfy the cargo demand only with a probability of at most 90%.

Therefore the version with stochastic demand for the vehicle routing problem can be written as follows:

$$\begin{aligned}\min : \text{Total cost} = & \ 10x_1 + 15x_2 + 18x_3 + 15x_4 + 32x_5 \\ & + 32x_6 + 57x_7 + 57x_8 + 60x_9 + 60x_{10} \\ & + 63x_{11} + 63x_{12} + 61x_{13} + 61x_{14} + 75x_{15} \\ & + 75x_{16} + 62x_{17} + 62x_{18} + 44x_{19},\end{aligned} \tag{4.22}$$

$$\text{s.t.} \quad \Pr\{\text{constraints}\,(4.7)\,\text{to}\,(4.20)\} \geq 0.9. \tag{4.23}$$

Eq. (4.23) indicates overall that the vehicles on each arc in the network can satisfy only 90% of the demand.

We can generalize the vehicle routing problem like this: There is a directed graph with a node set N and an arc set S. A set of cyclic routes Π has been selected, in which each route is a sequence of nodes connected with arcs and such that the last node of the sequence is the same as the first one. For each arc $s \in S$, we denote $R(s)$ as the set of routes containing arc s. For example, arc AB is contained by five routes: Route 1, Route 5, Route 11, Route 15, and Route 17, so $R(AB) = \{1, 5, 11, 15, 17\}$. For each route $\pi \in \Pi$, we denote $c(\pi)$ as the unit cost on route π.

Let $\xi(s)$ be the random integer demand of each arc $s \in S$, let $x(\pi)$ be the number of vehicles assigned on route π, and let α represent the confidence level. The objective is to find nonnegative integers of vehicles, $x(\pi), \pi \in \Pi$, such that

$$\Pr\left(\sum_{\pi \in R(s)} x(\pi)\, V_{\text{capacity}} \geq \xi(s), s \in S\right) \geq 1 - \alpha, \tag{4.24}$$

and the total cost

$$\sum_{\pi \in \Pi} c(\pi) x(\pi) \tag{4.25}$$

is minimized.

Therefore the chance-constrained programming model for the vehicle routing problem with uncertain demand on each arc is as follows:

$$\min \sum_{\pi \in \Pi} c(\pi) x(\pi), \tag{4.26}$$

$$\text{s.t.} \quad \Pr\left(\sum_{\pi \in R(s)} x(\pi) V_{\text{capacity}} \geq \xi(s), s \in S\right) \geq 1 - \alpha. \tag{4.27}$$

Finance Problem

Many stochastic programing models have been applied to financial problems. However, most of them that have been formulated so far belong to the class of recourse problems. The safety type considerations, such as the concept of value at risk and its variants, existed many years ago, but they have not gained enough attention during the past decades.

Now let us consider the classical portfolio models of finance problems proposed by Markowitz (1952, 1959, 1987). In his models, the safety aspects have been considered as the form of the variance of the return and incorporated into the models. Obviously, for any investor, the portfolio with a smaller variance of return is always preferred when its expected return is the same as others. Markowitz defined an efficient portfolio like this: Given the expectation of the return, its variance cannot be decreased and given the variance, and so the expectation cannot be increased. To illustrate the application of chance-constrained programming in finance problems, we present a bond portfolio construction model. The notations used in the model are as follows:

n, number of bond types, which are candidates for inclusion into the portfolio;

m, number of periods;

a_{ik}, cash flow of a bond of type k in period, $k=1,\ldots,n$, $i=1,\ldots,m$;

p_k, unit price of bond of type k;

ξ_i, random liability value in period i, $i=1,\ldots,m$;

x_k, decision variables, number of bonds of type k to include into the portfolio;

z_i, cash carried forward from period i to period $i+1$, $i=1,\ldots, m$, where z_1 is an initial cash amount that we include into the portfolio and $z_{m+1}=0; z_i$, $i=1,\ldots,m$ are decision variables; and

ρ_i, rate of interest in period i, $i=1,\ldots,m$.

First, we consider the liabilities as deterministic values, so that the optimal bond portfolio model would be the following:

$$\min\left\{\sum_{k=1}^{n} p_k x_k + z_1\right\}, \tag{4.28}$$

subject to

$$\sum_{k=1}^{n} a_{ik}x_k + (1-\rho_i)z_i - z_{i+1} \geq \xi_i, \quad i=1,\ldots,m \tag{4.29}$$

$$x_k \geq 0, \quad k=1,\ldots,n \tag{4.30}$$

$$z_i \geq 0, \quad i=1,\ldots,m, \; z_{m+1}=0. \tag{4.31}$$

Now we consider the liability as random values. Given a confidence level $\alpha \in (0, 1)$, then the probabilistic constrained variant of the model above can be reformulated as

$$\min\left\{\sum_{k=1}^{n} p_k x_k + z_1\right\}, \tag{4.32}$$

subject to

$$\Pr\left(\sum_{k=1}^{n} a_{ik}x_k + (1-\rho_i)z_i - z_{i+1} \geq \xi_i, i=1,\ldots,m\right) \geq 1-\alpha \tag{4.33}$$

$$x_k \geq 0, \quad k=1,\ldots,n \tag{4.34}$$

$$z_i \geq 0, \quad i=1,\ldots,m, \; z_{m+1}=0. \tag{4.35}$$

SUMMARY

This chapter presents three classical examples to introduce the mathematical modeling of chance-constrained programming problems. As can be seen from this chapter, the chance-constrained programming can be used in many areas, such as the classical vehicle routing problem, the finance problem, and water resource management problems, etc.

REFERENCES

Charnes, A., Cooper, W. W., & Symmonds, G. H. (1958). Cost horizons and certainty equivalents: an approach to stochastic programming of heating oil. *Management Science*, *4*, 235–263.

Dentcheva, D., Prékopa, A., & Ruszczyński, A. (2000). Concavity and efficient points of discrete distributions in probabilistic programming. *Mathematical Programming*, *89*(1), 55–77.

Dupacová, J., Gaivoronski, A., Kos, A., & Szántai, T. (1991). Stochastic programming in water management: A case study and a comparison of solution techniques. *European Journal of Operational Research*, *52*, 28–44.

Henrion, R., Li, P., Möller, A., Steinbach, S. G., Wendt, M., & Wozny, G. (2001). Stochastic optimization for operating chemical process under uncertainty. In M. Grötschel, S. Krunke, & J. Rambau (Eds.), *Online optimization of large scale systems* (pp. 457–478). Berlin: Springer.
Henrion, R., & Möller, A. (2003). Optimizatin of a continuous distillation process under random inflow rate. *Computer & Mathematics With Applications*, *45*, 247–262.
Luedtke, J., & Ahmed, S. (2008). A sample approximation approach for optimization with probabilistic constraints. *SIAM Journal of Optimization*, *19*, 674–699.
Markowitz, H. (1952). Portfolio selection. *Journal of Finance*, *7*, 77–91.
Markowitz, H. (1959). *Portfolio selection*. New York, NY: Wiley.
Markowitz, H. (1987). *Mean-variance analysis in portfolio choice and capital markets*. New York, NY: Blackwell.
Nemiroski, A., & Shapiro, A. (2006). Convex approximations of chance constrained programs. *SIAM Journal of Optimization*, *17*(4), 969–996.
Prékopa, A. (1995). *Stochastic programming*. Dordrecht: Kluwer Academic.
Prékopa, A., & Szántai, T. (1978a). A new multivariate gamma distribution and its fitting to empirical streamflow data. *Water Resources Research*, *14*, 19–24.
Prékopa, A., & Szántai, T. (1978b). Flood control reservoir system design using stochastic programming. *Mathematical Programming Study*, *9*, 138–151.

CHAPTER FIVE

Two-Stage Stochastic Model

Contents

INTRODUCTION

Chapter 4 introduced the modeling methodology of chance-constrained programming, which is one type of stochastic programming models. This chapter will further present another modeling methodology of stochastic programming, using some examples from a variety of areas to show its application, which is considered a two-stage stochastic programming model. Through these examples, readers will learn how to deal with the problem under uncertainty and how to build a model for it, as well as models for other problems with different structural aspects, such as different objectives of the decision process, the different forms of the constraints on those decisions, and the relationships to the random elements, etc.

In the first section, we reconsider the newsvendor problem presented in Chapter 3 and reformulate that problem from another point of view, as well as build a different stochastic programming model for it.

In the second section, we consider another example in which one can decide the areas of land that should be used for different plants. The yields of the crops will vary because the weather varies. From this example the

Liner Ship Fleet Planning
http://dx.doi.org/10.1016/B978-0-12-811502-2.00005-3

basic foundation of stochastic programming and the advantages of the stochastic programming solution over that of deterministic approach can be clearly illustrated to readers.

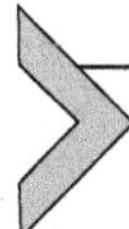

SELECTED APPLICATIONS

Newsvendor Problems

Let us reconsider the newsvendor problem, which is presented in Chapter 3. We can view the newsvendor problem as two stages. Let us analyze the problem in this way: In the morning, the newsvendor has to decide the quantity x of newspapers purchased from a distributor, but at that moment when he purchases from the distributor, he actually does not know the demand D of newspapers for that day. In other words the value of x has to be decided before the demand of newspapers D is known or realized. That's to say, x is made in the first stage. By the end of the day the value of D becomes known, so the newsvendor can optimize his behavior by selling as many newspapers as possible. This is to say that when the value of D is known, the amount of newspapers y at price s sold to customers, along with the amount of newspapers z at price r returned to distributors, would be decided after x is decided in the first stage; namely, y and z are decided in the second stage. Given a value of the first-stage decision variable x and a realization of the demand D, the newsvendor would encounter such a problem regarding how to decide y and z in order to maximize the profit of newspapers. The profit can be expressed as

$$\text{profit} = sy + rz. \tag{5.1}$$

Thus an optimization model would be developed to help the newsvendor make an optimal decision in the second stage:

$$\begin{aligned} \max_{y,z} \; & sy + rz \\ \text{s.t.} \;\; & y \leq D \\ & y + z \leq x \\ & y \geq 0 \\ & z \geq 0. \end{aligned} \tag{5.2}$$

The optimal solution of the problem (5.2) is $y^* = \min\{x, D\}$, $z^* = \max\{x - D, 0\}$, and its optimal value is the profit $F(x, D)$ defined in (3.1).

The above is a model to formulate the second-stage decision problem when x is given. However, x is the decision variable in the first stage. Therefore it is natural for us to consider the basic idea of two-stage decisions like this: At the first stage, before the random variables become known (i.e., before we know the value of D in the newsvendor problem), we have to choose the value for the first-stage decision variable x to optimize the expected value of an objective function shown in Eq. (5.1), which in turn is the optimal value of the second-stage optimization problem shown in Eq. (5.2). As in the first stage, the newsvendor has to purchase newspapers, thus the objective for the newsvendor in the first stage is to minimize his cost. While in the second stage, the newsvendor's objective is to maximize his profit of selling newspapers. In order to keep the consistency of the two objectives, we can write the two-stage stochastic programming model as follows:

$$\min\ cx + \mathbb{E}[Q(x, D)], \tag{5.3}$$

subject to

$$0 \leq x \leq M, \tag{5.4}$$

where

$$\begin{aligned} Q(x, D) = \min_{y, z}\ & -sy - rz \\ \text{s.t.}\ & y \leq D \\ & y + z \leq x \\ & y \geq 0 \\ & z \geq 0. \end{aligned} \tag{5.5}$$

The Farmer's Problem

Suppose a farmer has a land of 500 acres. He plans to plant some crops on this land, such as sugar beets, corn, and grain. For each winter, the farmer has to determine the acres of land for each crop. The farmer has cattle to feed, and according to experience, he has to keep at least 200 tons of wheat and 240 tons of corn in order to feed his cattle. This wheat and corn may be raised on the farm or bought from a wholesaler in the market. The farmer can sell the production if its yield is more than the feeding requirement in order to obtain additional revenue. The wheat can be sold at the price of \$170 per ton and the corn can be sold at the price of \$150 per ton. Considering

Table 5.1 Data for the Farmer's Problem

	Wheat	Corn	Sugar Beets
Yield (tons/acre)	2.5	3	20
Planting cost ($/acre)	150	230	260
Selling price ($/ton)	170	150	36 under 6000 tons 10 above 6000 tons
Purchase price ($/ton)	238	210	
Minimum requirement	200	240	

Total available land: 500 acres.

the transporting costs, the purchase prices should be higher than the selling price by at least 40%. As for sugar beets, there are two selling prices for it. This is because the European Commission imposes a quota on sugar beet production: When the amount of sugar is less than the quota, it can be sold at the $36/ton; otherwise, it could be sold at only $10/ton for the amount in excess of the quota. Here the farmer's quota for next year is 6000 tons. According to historical experience, the mean yield is 2.5 tons/acre for wheat, 3 tons/acre for corn, and 20 tons/acre for sugar beet. The data for the mean yield and for the planting costs for these crops are shown in Table 5.1.

The notations used in this problem can be listed as follows:

x_1 = acres of land devoted to wheat
x_2 = acres of land devoted to corn
x_3 = acres of land devoted to sugar beets
ω_1 = tons of wheat sold
y_1 = tons of wheat purchased
ω_2 = tons of corn sold
y_2 = tons of corn purchased
ω_3 = tons of sugar beets sold at the favorable price
ω_4 = tons of sugar beets sold at the lower price

For the farmer, the total cost includes planting and purchasing, which can be expressed as

$$\text{cost} = 150x_1 + 230x_2 + 260x_3 + 238y_1 + 210y_2. \quad (5.6)$$

The farmer can get the revenue by selling his crops, which can be expressed as

$$\text{revenue} = 170\omega_1 + 150\omega_2 + 36\omega_3 + 10\omega_4. \quad (5.7)$$

Therefore his profit can be computed as follows:

$$\begin{aligned} \text{profit} = {} & 170\omega_1 + 150\omega_2 + 36\omega_3 + 10\omega_4 \\ & - 150x_1 - 230x_2 - 260x_3 - 238y_1 - 210y_2. \end{aligned} \tag{5.8}$$

The first constraint is that the land needed for all crops cannot exceed the available area of land, namely

$$x_1 + x_2 + x_3 \le 500. \tag{5.9}$$

As at least 200 tons of wheat and 240 tons of corn are needed in order to feed his cattle, the following constraints hold:

$$2.5x_1 + y_1 - \omega_1 \ge 200 \tag{5.10}$$

$$3x_2 + y_2 - \omega_2 \ge 240. \tag{5.11}$$

The selling prices for sugar beets have two values prices (i.e., one is at a higher price and another is at lower price), so we have:

$$\omega_3 + \omega_4 \le 20x_3 \tag{5.12}$$

$$\omega_3 \le 6000. \tag{5.13}$$

Therefore if the farmer's aim is to maximize his profit while satisfying all the constraints, then the following optimization model holds:

$$\begin{aligned} \max 170\omega_1 + 150\omega_2 + 36\omega_3 + 10\omega_4 - 150x_1 - 230x_2 - 260x_3 \\ - 238y_1 - 210y_2, \end{aligned} \tag{5.14}$$

subject to

$$x_1 + x_2 + x_3 \le 500 \tag{5.15}$$

$$2.5x_1 + y_1 - \omega_1 \ge 200 \tag{5.16}$$

$$3x_2 + y_2 - \omega_2 \ge 240 \tag{5.17}$$

$$\omega_3 + \omega_4 \le 20x_3 \tag{5.18}$$

$$\omega_3 \le 6000 \tag{5.19}$$

$$x_1, x_2, x_3, y_1, y_2, \omega_1, \omega_2, \omega_3, \omega_4 \ge 0. \tag{5.20}$$

The above model is deterministic, and we can get an optimal solution by using a linear program solver, shown in Table 5.2.

If we carefully analyze the optimal solution shown in Table 5.2, then we can find that the optimal solution is easy to understand. The results show that the sugar beets have the most land devoted to their growth (i.e., compared with wheat and corn) in order to guarantee that the crop can reach the quota

Table 5.2 Optimal Solution Based on Expected Values of Yields

	Wheat	Corn	Sugar Beets
Yield (tons)	300	240	6000
Surface (acres)	120	80	300
Sales (tons)	100	–	6000
Purchase (tons)	–	–	–

Total profit: $118,600.

of 6000 tons. As for the remaining lands, they are used to plant wheat and corn in order to satisfy the feeding requirement. However the yields of crops are related to some uncontrollable and unpredictable factors such as weather, and therefore they are always varied; they cannot be predicted without error. According to past experience, yields varying 20 to 25 percent above or below the mean yield are not unusual. We have two different ways to deal with the representation of the yields: one approach uses discrete random variables and another uses continuous random variables.

A Scenario Representation

Based on the above analysis the weather will affect the yields of crops. In order to reflect the effects of weather on crop yields, we can simply assume that the years are good, fair, or bad for all crops. Correspondingly the yields of all crops have three scenarios: high, average, and low. The high scenario indicates that the yield is above average, while the low scenario indicates that the yield is below average. Here, in order to fix these ideas, "high" and "low" scenarios are assumed to indicate a yield 20% above or below the mean yield given in Table 5.1. Furthermore, we assume the prices of crops are still the same in all scenarios of yields for simplicity. Therefore the farmer can run two optimization models based on the yields corresponding to the "high" and "low" scenarios in order to get the corresponding optimal solutions; see Tables 5.3 and 5.4.

Table 5.3 Optimal Solution Based on High Scenario of Yields (+20%)

	Wheat	Corn	Sugar Beets
Yield (tons)	550	240	6000
Surface (acres)	183.33	66.67	250
Sales (tons)	350	–	6000
Purchase (tons)	–	–	–

Total profit: $167,667.

Table 5.4 Optimal Solution Based on Low Scenario of Yields (−20%)

	Wheat	Corn	Sugar Beets
Yield (tons)	200	60	6000
Surface (acres)	100	25	375
Sales (tons)	–	–	6000
Purchase (tons)	–	180	–

Total profit: $59,950.

We can analyze the optimal solutions again. The optimal solutions shown in Tables 5.3 and 5.4 seem to be almost the same as what we expected. When yields are high, less land is needed to meet the quota of sugar beets; conversely, more land is needed. For example, in Table 5.3, only 250 acres of land are needed to reach the quota; while in Table 5.4, 375 acres of land are needed. When yields are high, the remaining land is devoted to wheat, because its unit selling price is the highest; it can make the farmer earn more compared with corn or sugar beets. If we let the yields change, then we can find that the optimal solution changes as well. We can see that the optimal surfaces devoted to wheat range from 100 to 183.33 acres, the optimal surfaces devoted to corn range from 25 to 80 acres, and the optimal surfaces devoted to sugar beets range from 250 to 275 acres. The overall profit ranges from $59,950 to $167,667.

We find that the weather is quite important to the yields as well as to the optimal solutions, namely the assignment of acres of land and the resulting benefits. The farmer definitely hopes that the weather forecast for the long term is accurate so that he can make the best decisions. However, it is quite hard to accurately forecast the weather 6 months ahead; it is not easy to predict even for a few days in advance with any high accuracy. It is clear that the main issue here is sugar beet production because there are two selling prices related to the yields of sugar that would have two results: Large acres of land devoted to sugar beet production would ensure that the quote was met and sold, but it would also make it likely to have to sell some sugar beets at lower prices. However, by planting smaller surfaces, it would make it likely to miss the opportunity to sell the full quota at the favorable price.

Now we reconsider the problem from a different point of view. Obviously the realizations of yields of crops are behind the decision of land assignment. In other words, decisions on land assignment (x_1, x_2, x_3) have to be taken now, but sales and purchases $\left(\omega_i, i=1,\ldots,4; y_j, j=1,2\right)$ are determined later, depending on the yields. For the three scenarios of high, average, and low yields, we use the indexes by $s=1,2,3$ to denote these three scenarios, respectively. This creates news variables of the form $\omega_{is}, i=1,2,3,4, s=1,2,3$ and $y_{js}, j=1,2, s=1,2,3$.

Assuming that the farmer is neutral in regard to risk, and the three scenarios of yields occurred with an equal probability of 1/3, the farmer wants to seek a solution in order to maximize his expected profit. Consequently, we have the following model:

$$\begin{aligned} \max \frac{1}{3}&(170\omega_{11}+150\omega_{21}+36\omega_{31}+10\omega_{41}-238y_{11}-210y_{21}) \\ &+\frac{1}{3}(170\omega_{12}+150\omega_{22}+36\omega_{32}+10\omega_{42}-238y_{12}-210y_{22}) \\ &+\frac{1}{3}(170\omega_{13}+150\omega_{23}+36\omega_{33}+10\omega_{43}-238y_{13}-210y_{23}) \\ &-150x_1-230x_2-260x_3, \end{aligned} \tag{5.21}$$

subject to

$$x_1+x_2+x_3 \le 500 \tag{5.22}$$

$$3x_1+y_{11}-\omega_{11} \ge 200 \tag{5.23}$$

$$3.6x_2+y_{21}-\omega_{21} \ge 240 \tag{5.24}$$

$$\omega_{31}+\omega_{41} \le 24x_3 \tag{5.25}$$

$$\omega_{31} \le 6000 \tag{5.26}$$

$$2.5x_1+y_{12}-\omega_{12} \ge 200 \tag{5.27}$$

$$3x_2+y_{22}-\omega_{22} \ge 240 \tag{5.28}$$

$$\omega_{32}+\omega_{42} \le 20x_3 \tag{5.29}$$

$$\omega_{32} \le 6000 \tag{5.30}$$

$$2x_1+y_{13}-\omega_{13} \ge 200 \tag{5.31}$$

$$2.4x_2+y_{23}-\omega_{23} \ge 240 \tag{5.32}$$

$$\omega_{33}+\omega_{43} \le 16x_3 \tag{5.33}$$

$$\omega_{33} \le 6000 \tag{5.34}$$

$$x_k, y_{js}, \omega_{is} \ge 0 \quad (k=1,2,3;\ i=1,2,3,4;\ j=1,2;\ s=1,2,3). \tag{5.35}$$

The second-stage decision variables for all scenarios are explicitly described by this model. We can solve this model and show its optimal solution in Table 5.5. The top line gives the planting areas, which must be decided before realizing the weather and crop yields, so it is considered the first-stage decision variables. The other lines describe the yields, sales, and purchases in the three scenarios, which are the second-stage

Table 5.5 Optimal Solutions for All Scenarios of Yields

		Wheat	**Corn**	**Sugar Beets**
First stage	Area (acres)	170	80	250
$s=1$	Yield (T)	510	288	6000
	Sales (T)	310	48	6000
	Purchase (T)	–	–	–
$s=2$	Yield (T)	425	240	5000
	Sales (T)	225	–	5000
	Purchase (T)	–	–	–
$s=3$	Yield (T)	340	192	4000
	Sales (T)	140	–	4000
	Purchase (T)	–	48	–

Overall profit: $108,390.

decision variables. The bottom line is the overall expected profit earned by the farmer.

We can explain the optimal solution as follows: in order to make the most profitable decision for sugar beet land allocation, we should always avoid sales at the lower price, even if this implies that some portion of the quota is unused when yields are average or below average. As for corn, we can find that the devoted land area is such that it meets the feeding requirement when the yield is average, which implies that when yield is high, it is possible for the farmer to sell excess corn. However, when yield is low, the farmer may need to purchase some corn in order to feed his cattle. As for the rest of the land, it is all used for planting wheat.

The solutions shown in Tables 5.4 and 5.5 illustrate that it is impossible to find a solution that is ideal under all scenarios of uncertain yields. If the farmer could predict the yields of his plants precisely, then he can make a best plan for his plants in order to maximize his revenue. As in a stochastic model, various scenarios are always encountered by decision makers, so decisions have to be balanced for these different scenarios.

If we suppose that the farmer's crop yields vary cyclically every three years, for example, a year with high yields may be followed by a year with average yields, then by a year with low yields. The farmer would then get profits shown in Tables 5.2–5.4, respectively; that is, he would get profit of $167,667 the first year, $118,600 the second year, and $ 59,950 the third year. Therefore the mean profit over the 3 years would be equal to (167,667 + 118,600 + 59,950)/3 = $115,406.

The yields of these plants are assumed to be various and uncertain over the years. If the information on yields is known by the farmer before he plants crops, he would again assign his land based on the solution shown in Tables 5.2–5.4. However, it is impossible for the farmer to get prior information on the yields, so he has to assign his land based on the solution shown in Table 5.5 and get an expected profit of $108,390. The difference between $115,406 and $108,390 is $7016, which is called *the expected value of perfect information* (EVPI); it represents the loss of the profit due to the presence of uncertainty.

Another approach for the farmer to adopt is to allocate his land based on the case of the expected yields, as shown in Table 5.2. Mathematically, the solution in Table 5.2 is called the *expected value solution*. Now the difference between the profit of $118,600 and $108,390, which is $10,210, is called the *value of the stochastic solution* (VSS); this is the possible gain from solving the stochastic model, which is larger than the EVPI.

According to the concepts of EVPI and VSS, they are slightly different: EVPI represents the difference of the average value of optimal solutions to all models of all scenarios and the optimal solution to the expected value model, while VSS is to assess the difference of the optimal solution to the model in which uncertain parameter is replaced by its mean value and the optimal solution to the expected value model. Therefore, if there is no more information about uncertainty in future is available for a problem, the VSS is more practical than EVPI. In other situations, more information may be available through more extensive forecasting, sampling, or exploring, in these cases, EVPI would be useful for deciding whether to undertake additional efforts.

General Model Formulation

The example of the farmer's problem described above can be used to illustrate the general formulation of a two-stage stochastic programming model. Through the example, we can see that there is a set of decisions to be taken without full information on some random event (like the assignment of lands to crops has to be made before knowing the yields). These decisions are known as *first-stage decisions* and are usually denoted by a vector variable x. Later the random events occurred and the information is obtained. The random events can be represented by a vector ξ, and a realization is denoted by ω or s. After the random events occurred and the information is realized, then corrective actions are taken, which are called *second-stage decisions* and are usually represented by a vector y. Let take the farmer example to demonstrate the random events and the corrective actions: Obviously, the yields of plants are random events because they are related to the uncontrollable and unpredictable factors; the purchase and sales of products are corrective actions. Mathematically, we can define the problem as a two-stage stochastic program as follows:

$$\min c^T x + \mathbb{E}_\xi[Q(x, \xi)] \tag{5.36}$$

$$\text{s.t. } Ax = b, \; x \geq 0, \tag{5.37}$$

where $Q(x, \xi) = \min\{q^T y | Wy = h - Bx, y \geq 0\}$, and $\mathbb{E}_\xi$ denotes mathematical expectation with respect to ξ.

In the farmer example the random vector is a discrete variable with only three different scenarios: high, average, and low. For one specific scenario s, it can be formulated as a *second-stage problem* as follows:

$$\begin{aligned} Q(x, s) = \min\ & (-170\omega_1 - 150\omega_2 - 36\omega_3 - 10\omega_4 + 238y_1 + 210y_2) \\ \text{s.t. } & b_1(s)x_1 + y_1 - \omega_1 \geq 200, \\ & b_2(s)x_2 + y_2 - \omega_2 \geq 240, \\ & \omega_3 + \omega_4 \leq b_3(s)x_3, \\ & \omega_3 \leq 6000, \\ & y, \omega \geq 0, \end{aligned} \tag{5.38}$$

where $t_i(s)$ $(i = 1,2,3)$ denotes the yield of crops i $(i = 1,2,3)$ under scenario s. Therefore for all of three scenarios, the expected value for the second-stage problem would be:

$$\begin{aligned} & \mathbb{E}_\xi[Q(x, \xi(s))] \\ & = \mathbb{E}_\xi\left[\min\begin{pmatrix} 238y_1(\xi(s)) + 210y_2(\xi(s)) \\ -170\omega_1(\xi(s)) - 150\omega_2(\xi(s)) - 36\omega_3(\xi(s)) - 10\omega_4(\xi(s)) \end{pmatrix}\right] \\ & \text{s.t. } b_1(\xi(s))x_1 + y_1(\xi(s)) - \omega_1(\xi(s)) \geq 200, \\ & \qquad b_2(\xi(s))x_2 + y_2(\xi(s)) - \omega_2(\xi(s)) \geq 240, \\ & \qquad \omega_3(\xi(s)) + \omega_4(\xi(s)) \leq b_3(\xi(s))x_3, \\ & \qquad \omega_3(\xi(s)) \leq 6000, \\ & \qquad y, \omega \geq 0. \end{aligned} \tag{5.39}$$

In the farmer example the first-stage decisions include the allocation of lands to wheat, corn, and sugar beets. Therefore in the first stage, the costs refer to planting the three crops can be expressed as $150x_1 + 230x_2 + 260x_3$; obviously, $x_1 + x_2 + x_3 \leq 500$ is the constraint in the first stage. It is a common sense for the farmer to minimize the total costs, which are the sum of the planting costs of the first stage and the expected costs in the second stage. Therefore we can propose a two-stage stochastic programming model as follows:

$$\begin{aligned} \min\ & 150x_1 + 230x_2 + 260x_3 + \mathbb{E}_\xi[Q(x, \xi(s))] \\ \text{s.t.}\ & x_1 + x_2 + x_3 \leq 500 \\ & x_1, x_2, x_3 \geq 0 \end{aligned} \tag{5.40}$$

where $\mathbb{E}_\xi[Q(x, \xi(s))]$ is formulated in Eq. (5.39).

It is noted that in our example, the distribution ξ of random events is assumed to be discrete and represented by three scenarios, but in the general form, it can be continuous as well.

SUMMARY

This chapter presents three classical examples to introduce the mathematical modeling of two-stage stochastic programming problems. As can be seen from this chapter the two-stage stochastic programming can be used in many areas; it is not easy to solve because the second-stage optimization problem is merged into the first stage. For the solving method, we will talk about it in detail in Part III of this book.

FURTHER READING

Birge, John R., & Francois, Louveaux (1997). *Introduction to stochastic programing*. New York: Springer Press.

Ruszczyński, A., & Shapiro, A. (2003). *Handbooks in operational research and management science*. (Vol. 10). North Holland: Elsevier Science Press.

PART THREE

Solution Algorithms

CHAPTER SIX

Sample Average Approximation

Contents

INTRODUCTION

This chapter introduces the modeling methodology of chance-constrained programming. Problems with probabilistic constraints (i.e., chance constraints) have attracted much of the attention of researchers for a long time (see Prékopa, 2003 for the basic form, theorem, and applications). Chance constraints are used in many problems under uncertain environments, including scheduling problems (Henrion et al., 2001; Henrion & Möller, 2003), production planning (Murr & Prékopa, 2000), supply chain management (Lejeune & Ruszczyński, 2007), and water management (Takyi & Lence, 1999).

This chapter will introduce the solution algorithm used to solve chance-constrained problems. Generally, we can formulate the chance-constrained problems as the following optimization model:

$$\begin{aligned} &\min_{x \in X} f(x) \\ \text{s.t.}\quad &\text{prob}\{G(x, \xi) \leq 0\} \geq 1 - \alpha \end{aligned} \tag{6.1}$$

where $X \subset \mathbb{R}^n$, ξ is a random vector with a probability distribution P on a set $\Xi \subset \mathbb{R}^d, \alpha \in (0, 1), f: \mathbb{R}^n \to \mathbb{R}$ is a function mapping between real sets, and $G: \mathbb{R}^n \times \Xi \to \mathbb{R}^m$.

Liner Ship Fleet Planning
http://dx.doi.org/10.1016/B978-0-12-811502-2.00006-5

It has been over 50 years since chance-constrained programming was proposed by Charnes, Cooper, & Symmonds, 1958, but little progress was made in methods to solve it until recently. There are two reasons that the chance-constrained programming model is quite difficult to solve. Even though the function $G(x, \xi)$ is quite simple, it is still hard to solve numerically. One reason for this is that the computation of the value of $\text{prob}\{G(x, \xi) \leq 0\}$ for a given $x \in X$ is not easy for us, because we need to calculate a multidimensional integration as ξ is a random vector. Consequently, a possible and treatable way is to use Monte Carlo simulation to check whether a given $x \in X$ is feasible.

Another reason is that we cannot guarantee that the feasible set of problem (6.1) must be always convex; even if the set X is convex and the function $G(x, \xi)$ is convex in x, it could still possibly be nonconvex. There is currently no good solution to solve an optimization problem with a nonconvex feasible set. Therefore, we can divide the solving methodologies into two somewhat different directions: One is to discretize the probability distribution P, then solve the obtained problems (see, e.g., Dentcheva, Prékopa, & Ruszczyński, 2000; Luedtke & Ahmed, 2008). Another approach is to approximate the chance constraints by employing convex approximations (e.g., Nemiroski & Shapiro, 2006).

In this chapter, we introduce the sample average approximation (SAA) method to solve chance-constrained problems, which is one of the methods belonging to the first direction. The SAA method is to replace the probability or actual distribution in chance constraints by the frequency or empirical distribution corresponding to a random sample. This method has been investigated by Luedtke and Ahmed (2008) and Atlason, Epelman, and Henderson (2008), and it will be discussed in detail in the following sections.

THEORETICAL BACKGROUND

We have pointed out that the difficulty to solve the chance-constrained problem is the issue of how to obtain the probability in the constraints. Furthermore, there is also the issue of dealing with the probabilistic form of constraints. Generally, it's not easy for us to obtain a closed form of the probability and its distribution in the constraints; even if we can get it, the probabilistic form of constraints is still hard to tackle. As the closed form of probability is quite hard for us to obtain, we use an approximation to replace it. Alhough it is not a perfect method due to some gap between the approximation and the real probability, it is at least a feasible method for us to use. In order to approximate the probability, we can use frequency to replace it, because the probability of an event can be simply explained as the frequency of the occurrence of the event in a number of tests.

Let us revisit the chance constraints in Eq. (6.1). Firstly, we assume that the constraints function $G:\mathbb{R}^n \times \Xi \rightarrow \mathbb{R}$ is mapping between real sets, namely the region of the objective value of function is a real set, in order to simplify the statement. If we define

$$p(x) := P\{G(x, \xi) > 0\}, \tag{6.2}$$

then we have the following equivalent form for problem (6.1):

$$\min_{x \in X} f(x) \quad \text{s.t. } p(x) \leq \alpha. \tag{6.3}$$

It is noted that the transformation technique can still be used for the case of joint distribution. For example, if there are m constraints $G_i(x, \xi) \leq 0, \quad i = 1, \ldots, m$, it can be replaced equivalently by one constraint

$$G(x, \xi) := \max_{1 \leq i \leq m} G_i(x, \xi) \leq 0. \tag{6.4}$$

Now if $\xi^1, \ldots, \xi^N$ is an independent identically distributed (iid) sample of N realizations of random vector ξ, then the probability $p(x)$ can be approximately replaced by the proportion of times that $G(x, \xi) > 0$ accounted for the N samples. In order to indicate the constraint $G(x, \xi)$, which is larger than zero, we define the indicator function of $(0, \infty)$; that is,

$$1_{(0,\infty)}(t) := \begin{cases} 1, & \text{if } t > 0 \\ 0, & \text{if } t \leq 0 \end{cases}, \tag{6.5}$$

and define the SAA of probability $p(x)$ to be $\hat{p}_N(x)$:

$$\hat{p}_N(x) := P_N\{G(x, \xi) > 0\}. \tag{6.6}$$

Then we can obtain the approximation of $p(x)$:

$$p(x) \approx \hat{p}_N(x) = \frac{\sum_{i=1}^{N} 1_{(0,\infty)}\left(G(x, \xi^i) > 0\right):}{N}, \tag{6.7}$$

It is noted that for any x, $\hat{p}_N(x) \xrightarrow{w.p.1} p(x)$ by strong Large Numbers Law. Therefore, we have such a problem associated with the generated sample $\xi^1, \ldots, \xi^N$:

$$\min_{x \in X} f(x) \quad \text{s.t. } \hat{p}_N(x) \leq \gamma \tag{6.8}$$

The problem (6.3) is known as the true problem at the significance level α. Correspondingly the problem (6.8) is known as the SAA problem at the significance level γ. Here, the value to the significance level γ can be different from that to the significant level α, because when the SAA problem with significant level γ would yield lower/upper bounds to the true problem, when the SAA problem with significant level γ which is equal to α, the solution of the SAA problem can be convergent to that of the true problem with respect to the sample size N and the significant level γ, which has been proved by Pagnoncelli, Ahmed, and Shapiro (2009). The following section briefly presents the complementary convergence analysis of the SAA problem.

CONVERGENCE

If the function $G(x, \xi)$ is a Carathéodory function, then the functions $p(x)$ and $\hat{p}_N(x)$ are both lower semicontinuous. Based on the lower semicontinuity of $p(x)$ and $\hat{p}_N(x)$, we can obtain such a conclusion that the true problem (6.3) and its SAA counterpart (6.8) both have closed feasible sets. Further, if the set X is compact and the feasible sets for the problems (6.3) and (6.8) are nonempty, then the sets of optimal solutions denoted by S and $\hat{S}_N$, respectively, contain at least one solution; namely S and $\hat{S}_N$ are both nonempty as well. Let ϑ^* and $\hat{\vartheta}_N$ be the values of the optimal solutions to the true problem (6.3) and the SAA problem (6.8), respectively. Then we can show that for the case in which $\gamma = \alpha$, $\hat{\vartheta}_N$ and $\hat{S}_N$ of the SAA problem converge with probability 1 to ϑ^* and S of the true problem.

Assumption 6.1. There exists a sequence $\{x_k | p(x_k) \leq \alpha\} \subset X$ that converges to an optimal solution $\bar{x} \in S$; that is, $\bar{x}$ is an accumulation point of the set $\{x \in X | p(x) \leq \alpha\}$.

Proposition 6.1. Assuming that the domains of the true problem (6.3) and the SAA problem (6.8) are both compact, the function $f(x)$ is continuous, and $G(x, \xi)$ is a Carathéodory function, the significance level of the true problem (6.3) and SAA problem (6.8) are the same; that is, $\gamma = \alpha$, and so the assumption 1 holds. Then $\hat{\vartheta}_N \to \vartheta^*$ and $\mathbb{D}(\hat{S}_N, S) \to 0$ with probability 1 as $N \to \infty$.

Proof. By Assumption 6.1, the set S is nonempty, and there is $x \in X$ such that $p(x) \leq \alpha$. As $\hat{p}_N(x)$ converges to $p(x)$ with probability 1, consequently we have $\hat{p}_N(x) \leq \alpha$, and then the SAA problem has a feasible solution with probability 1 for sufficiently large N. By the lower semicontinuity of

$\hat{p}_N(x)$, the SAA problem has a compact feasible set; hence $\hat{S}_N$ is nonempty with probability 1 for sufficient large N. If x is a feasible solution of an SAA problem, then $f(x) \geq \hat{\vartheta}_N$. According to Assumption 6.1, we can take x arbitrarily close to $\bar{x}$ to obtain

$$\limsup_{N\to\infty} \hat{\vartheta}_N \leq f(\bar{x}) = \vartheta^*, \quad \text{w.p.1.} \tag{6.9}$$

Now let $\hat{x}_N \in \hat{S}_N$; that is, $\hat{x}_N \in X, \hat{p}_N(\hat{x}_N) \leq \alpha$ and $\hat{\vartheta}_N = f(\hat{x}_N)$. As the set X is compact, there exists a subsequence that $\hat{x}_N$ converges to a point $\bar{x} \in X$ w.p.1; hence $\liminf_{N\to\infty} \hat{p}_N(\hat{x}_N) \geq p(\bar{x})$ w.p.1. As $p(\bar{x}) \leq \alpha$ and therefore $\bar{x}$ is contained in the feasible set of the true problem, the result is $f(\bar{x}) \geq \vartheta^*$. Also, $f(\hat{x}_N) \to f(\bar{x})$ w.p.1, and hence

$$\liminf_{N\to\infty} \hat{\vartheta}_N \geq \vartheta^*, \quad \text{w.p.1.} \tag{6.10}$$

It follows from Eqs. (6.9) and (6.10) that $\hat{\vartheta}_N \to \vartheta^*$ w.p.1. We can also obtain that the point $\bar{x}$ is an optimal solution to the true problem, and $\mathbb{D}(\hat{S}_N, S) \to 0$ with probability 1 as $N \to \infty$. The proposition has been proven.□

LOWER BOUND

The solution of the SAA problem could converge to the solution of the true problem, but such convergence requires mild regularity conditions, which has been proved in the above section. If those conditions are unsatisfied, the convergence cannot be guaranteed. However, we can obtain the lower bounds to the true problem through the SAA problem even without those conditions.

We now establish that the SAA problem (6.8) yields a lower bound to the true problem (6.3) on a probability. Let

$$B(k; \alpha, N) = \sum_{i=0}^{k} \binom{N}{i} \alpha^i (1-\alpha)^{N-i} \tag{6.11}$$

It denotes the probability of an event in which there are at most k times of "successes" in N independent trials; in each trial the probability of a success is α.

Lemma 6.1. Assuming there is an optimal solution to the true problem (6.3), then

$$\Pr\{\hat{\vartheta}_N \leq \vartheta^*\} \geq B(\lfloor \gamma N \rfloor; \alpha, N). \tag{6.12}$$

Proof. Let $X_\alpha = \{x \in X : p(x) \leq \alpha\}$ denote the set of the feasible solution to the true problem (6.3), and let $X_\gamma^N = \{x \in X : \hat{p}_N(x) \leq \gamma\}$ denote the set of feasible solution to the corresponding SAA problem (6.8). If $x^* \in S$ is an optimal solution to the true problem (6.3), then we have $P\{G(x^*, \xi^i) > 0\} \leq \alpha$ for the ith trial. Let us call the event $\{G(x^*, \xi^i) > 0\}$ as a success, then the probability of a success in the ith trial can be calculated by the equation: $\overline{\phi}(x^*) := P\{G(x^*, \xi^i) > 0\} \leq \alpha$. By definition of X_γ^N, $x^* \in X_\gamma^N$ if and only if the following hold:

$$\frac{\sum_{i=1}^{N} 1_{(0,\infty)}\left(G(x^*, \xi^i) > 0\right)}{N} \leq \gamma \Leftrightarrow \sum_{i=1}^{N} 1_{(0,\infty)}\left(G(x^*, \xi^i) > 0\right) \leq \lfloor N\gamma \rfloor. \tag{6.13}$$

Hence $\Pr\{x^* \in X_\gamma^N\}$ is the probability of having at most $\lfloor N\gamma \rfloor$ successes in N trials; that is,

$$\Pr\{x^* \in X_\gamma^N\} = B(\lfloor \gamma N \rfloor; \overline{\phi}(x^*), N) \geq B(\lfloor \gamma N \rfloor; \alpha, N). \tag{6.14}$$

Also, if $x^* \in X_\gamma^N$, then $\hat{\vartheta}_N \leq \vartheta^*$, and so we have

$$\Pr\{\hat{\vartheta}_N \leq \vartheta^*\} \geq \Pr\{x^* \subset X_\gamma^N\} = B(\lfloor \gamma N \rfloor; \overline{\phi}(x^*), N) \geq B(\lfloor \gamma N \rfloor; \alpha, N). \tag{6.15}$$

Lemma 6.1 has been proven.□

Now we suppose that $\gamma > \alpha$, and we can find that the SAA problem would yield a lower bound with probability approaching 1 exponentially fast as the size N increases. Such a founding is based on Hoeffding's inequality (Hoeffding, 1963).

Theorem 6.1 (Hoeffding's Inequality). Let $Y_1, \ldots, Y_N$ be independent random variables with $\Pr\{Y_i \in [a_i, b_i]\} = 1$, where $a_i \leq b_i$ for $i = 1, \ldots, N$. If $t > 0$, the following inequality then holds:

$$\Pr\left\{\sum_{i=1}^{N} (Y_i - \mathbb{E}[Y_i]) \geq tN\right\} \leq \exp\left\{-\frac{2N^2t^2}{\sum_{i=1}^{N}(b_i - a_i)^2}\right\}. \tag{6.16}$$

Theorem 6.2. Let $\gamma > \alpha$, and suppose that there is an optimal solution to the true problem (6.3). Then,

$$\Pr\{\hat{\vartheta}_N \leq \vartheta^*\} \geq 1 - \exp\{-2N(\lambda - \alpha)^2\}. \tag{6.17}$$

Proof. Let $x^* \in S$ be an optimal solution to the true problem (6.3). As in the proof of Lemma 6.1, if $x^* \in X_\gamma^N$, then $\hat{\vartheta}_N \leq \vartheta^*$. For $i = 1, \ldots, N$, let Y_i be a random binary variable that equals 1 if $G(x^*, \xi^i) > 0$; otherwise, it will be 0. Then, $\Pr\{Y_i \in [0, 1]\} = 1$ and $\mathbb{E}[Y_i] \leq \alpha$. Hence

$$\begin{aligned}
\Pr\{\hat{\vartheta}_N > \vartheta^*\} &\leq \Pr\{x^* \notin X_\gamma^N\} = \Pr\left\{\frac{1}{N}\sum_{i=1}^{N} Y_i > \gamma\right\} \\
&\leq \Pr\left\{\frac{1}{N}\sum_{i=1}^{N}(Y_i - \mathbb{E}[Y_i]) > \gamma - \alpha\right\} \\
&\leq \exp\left\{-\frac{2N^2(\gamma-\alpha)^2}{N}\right\} = \exp\{-2N(\gamma-\alpha)^2\},
\end{aligned} \tag{6.18}$$

where the first inequality follows, as $\mathbb{E}[Y_i] \leq \alpha$, and the second inequality follows from Hoeffding's Inequality. Theorem 6.1 has been proven.□

Theorem 6.2 implies the method to obtain a lower bound to the value of the optimal solution of the true problem (6.3): we just need to take an parameter $\gamma > \alpha$ in the SAA problem (6.8). It has to be noted that the probability with which we can obtain a lower bound is approaching 1 exponentially fast as N increases. In addition, Theorem 6.2 yields a method to generate lower bounds with a given confidence level $1 - \delta$. If we select $\gamma > \alpha$ and let $N \geq \log(1/\delta)/(2(\gamma - \alpha)^2)$, we have

$$\Pr\{\hat{\vartheta}_N > \vartheta^*\} \leq \exp\{-2N(\gamma - \alpha)^2\} \leq \exp\{-\log(1/\delta)\} = \delta. \tag{6.19}$$

That is, Theorem 6.2 ensures that $\hat{\vartheta}_N \leq \vartheta^*$ with probability at least $1 - \delta$.

FEASIBLE SOLUTIONS

Theorems 6.1 and 6.6 show that the SAA problem (6.8) can yield a lower bound to the true problem (6.3) with a probability approaching 1 exponentially fast as N increases by taking a parameter $\gamma > \alpha$ in the SAA problem. Now we have determine whether the solutions of the SAA problem are feasible for the true problem as well; namely, if an optimal solution to the SAA problem exists, denoted by x, and whether $x \in X_\alpha$. If we let $\gamma < \alpha$, then for N large enough, the feasible solution set X_γ^N will be a subset of the feasible region of the true problem, so that any optimal solution $x \in \hat{S}_N$ must result in $x \in X_\alpha$. Now we discuss the assumptions to ensure the SAA problem yields a feasible solution to the true problem with a high probability.

Theorem 6.3. Suppose X is finite and $\gamma \in [0, \alpha)$, then

$$\Pr\left\{X_\gamma^N \subseteq X_\alpha\right\} \geq 1 - |X \backslash X_\alpha| \exp\left\{-2N(\alpha-\gamma)^2\right\}. \tag{6.20}$$

Proof. Consider any $\tilde{x} \in X \backslash X_\alpha$, that is, $\tilde{x} \in X$ with $\Pr\{G(\tilde{x}, \xi) \leq 0\} < 1-\alpha$. For $i = 1, \ldots, N$, let Y_i be a random binary variable that equals 1 if $G(\tilde{x}, \xi^i) \leq 0$ but equals 0 otherwise. Then $\Pr\{Y_i \in [0, 1]\} = 1$ and $\mathbb{E}[Y_i] = \Pr\{G(\tilde{x}, \xi^i) \leq 0\} < 1-\alpha$. As we have $\tilde{x} \in X_\gamma^N$ if and only if $(1/N)\sum_{i=1}^N Y_i \geq 1-\gamma$, and by using Hoeffding's inequality, we have

$$\begin{aligned}\Pr\left\{\tilde{x} \in X_\gamma^N\right\} = \Pr\left\{\frac{1}{N}\sum\nolimits_{i=1}^N Y_i \geq 1-\gamma\right\} &\leq \Pr\left\{\sum\nolimits_{i=1}^N (Y_i - \mathbb{E}[Y_i]) \geq N(\alpha-\gamma)\right\}. \\ &\leq \exp\left\{-2N(\alpha-\gamma)^2\right\}\end{aligned} \tag{6.21}$$

Then

$$\begin{aligned}\Pr\left\{X_\gamma^N \not\subset X_\alpha\right\} &= \Pr\left\{\exists \tilde{x} \in X_\gamma^N \text{ s.t. } \Pr\{G(\tilde{x}, \xi) \leq 0\} < 1-\alpha\right\} \\ &\leq \sum_{\tilde{x} \in X \backslash X_\alpha} \Pr\left\{\tilde{x} \in X_\gamma^N\right\} \\ &\leq |X \backslash X_\alpha| \exp\left\{-2N(\alpha-\gamma)^2\right\}.\end{aligned} \tag{6.22}$$

Theorem 6.3 yields a method to generate a feasible solution to the true problem (6.3) with a given confidence level $1-\delta$. If we select $\gamma < \alpha$ and let $N \geq \frac{1}{2(\alpha-\gamma)^2} \log\left(\frac{|X \backslash X_\alpha|}{\delta}\right)$, we have

$$\Pr\left\{X_\gamma^N \subseteq X_\alpha\right\} \geq 1 - |X \backslash X_\alpha| \exp\left\{-2N(\alpha-\gamma)^2\right\} \geq 1-\delta. \tag{6.23}$$

That is, Theorem 6.3 ensures that $X_\gamma^N \subseteq X_\alpha$ with a probability of at least $1-\delta$.

Theorem 6.3 states that for $\gamma < \alpha$, every feasible solution to the SAA problem will also be feasible to the true problem with a high probability as N gets large, which indicates that we need not solve the SAA problem to optimality in order to obtain a solution to the true problem. More importantly, if we can get a feasible solution to the true problem, it obviously yields an upper bound to the true problem.□

EXAMPLES

Let us consider an optimization problem as follows:

$$\min_{x_1 \geq 0,\, x_2 \geq 0} x_1 + x_2 \text{ s.t. } \Pr\{\xi_1 x_1 + x_2 \geq 7, \xi_2 x_1 + x_2 \geq 4\} \geq 1 - \alpha, \qquad (6.24)$$

where $x_i, i = 1, 2$ are the decision variables and $\alpha \in [0, 1]$ is the confidence level, ξ_1 and ξ_2 are independent continuous uniform random variables distributed on the intervals [1, 4] and [1/3, 1], respectively.

In order to solve the problem (6.24), we need to deal with the joint chance constrained in the problem using the min (or max) operators. Therefore the problem (6.24) becomes

$$\min_{x_1 \geq 0,\, x_2 \geq 0} x_1 + x_2 \text{ s.t. } \Pr\{\min\{\xi_1 x_1 + x_2 \geq 7, \xi_2 x_1 + x_2 \geq 4\}\} \geq 1 - \alpha. \quad (6.25)$$

We can rewrite the problem (6.25) in the following SAA form by introducing auxiliary binaries $z^i (i = 1, \ldots, N)$

$$\begin{aligned} &\min_{x_1 \geq 0,\, x_2 \geq 0} x_1 + x_2 \\ &\text{s.t. } \xi_1^i x_1 + x_2 - 7 + K z^i \geq 0 \quad i = 1, \ldots, N \\ &\qquad \xi_2^i x_1 + x_2 - 4 + K z^i \geq 0 \quad i = 1, \ldots, N \\ &\qquad \sum_{i=1}^{N} z^i \leq N\gamma \qquad\qquad i = 1, \ldots, N, \end{aligned} \qquad (6.26)$$

where N is the number of samples, ξ_1^i and ξ_2^i are realizations from ξ_1 and ξ_2, $\gamma \in (0, 1)$, and K is a positive constant not less than 7.

SUMMARY

This chapter presents the SAA method to solve the chance-constrained programming model. We have shown that the solution of the SAA problem can be convergent to the solution of the true problem under mild conditions, and the SAA problem can provide the lower and upper bound to the true problem. In the next chapter, we will present the algorithm for solving the two-stage stochastic programming model.

REFERENCES

Atlason, J., Epelman, M. A., & Henderson, S. G. (2008). Optimizing call center staffing using simulation and analytic center cutting plane methods. *Management Science, 54*, 295–309.

Charnes, A., Cooper, W. W., & Symmonds, G. H. (1958). Cost horizons and certainty equivalents: an approach to stochastic programming of heating oil. *Management Science, 4*, 235–263.

Dentcheva, D., Prékopa, A., & Ruszczyński, A. (2000). Concavity and efficient points of discrete distributions in probabilistic programming. *Mathematical Programming, 89*(1), 55–77.

Henrion, R., Li, P., Möller, A., Steinbach, S. G., Wendt, M., & Wozny, G. (2001). Stochastic optimization for operating chemical process under uncertainty. In M. Grötschel, S. Krunke, & J. Rambau (Eds.), *Online optimization of large scale systems* (pp. 457–478). Berlin: Springer.

Henrion, R., & Möller, A. (2003). Optimizatin of a continuous distillation process under random inflow rate. *Computer and Mathematics with Applications, 45*, 247–262.

Hoeffding, A. (1963). Probability inequalities for sums of bounded random variables. *Journal of American Statistics Association, 58*, 13–30.

Lejeune, A. A., & Ruszczyński, A. (2007). An efficient trajectory method for probabilistic production-inventory-distribution problems. *Operations Research, 55*(2), 378–394.

Luedtke, J., & Ahmed, S. (2008). A sample approximation approach for optimization with probabilistic constraints. *SIAM Journal of Optimization, 19*, 674–699.

Murr, A. R., & Prékopa, A. (2000). Solution of a product substitution problem using stochastic programming. In S. P. Uryasev (Ed.), *Probabilistic constrained optimization: methodology and applications* (pp. 252–271). North Holland: Kluwer Academic Publisher.

Nemiroski, A., & Shapiro, A. (2006). Convex approximations of chance constrained programs. *SIAM Journal of Optimization, 17*(4), 969–996.

Pagnoncelli, B. K., Ahmed, S., & Shapiro, A. (2009). Sample average approximation method for chance constrained programming: theory and applications. *Journal of Optimization Theory and Application, 142*, 399–416.

Prékopa, A. (2003). Probabilistic programming. In A. Ruszczyński & A. Shapiro (Eds.), *Vol. 10. Handbooks in operational research & management science* (pp. 267–351). North Holland: Elsevier Science Press.

Takyi, A. K., & Lence, B. J. (1999). Surface water quality management using a multiple-realization chance constraint method. *Water Resource Research, 35*, 1657–1670.

FURTHER READING

Chernoff, H. (1952). A measure of asymptotic efficiency for tests of a hypothesis based on the sum observations. *Annals of Mathematical Statistics, 23*, 493–507.

CHAPTER SEVEN

Dual Decomposition and Lagrangian Relaxation

Contents

INTRODUCTION

This chapter introduces the modeling methodology of two-stage stochastic programming. In the two-stage stochastic programming the first-stage decision variables can be regarded as here-and-now decisions, while the second-stage decision variables can be regarded as wait-and-see decisions. The second-stage decision variables are made after the first-stage decisions have been made. The aim of the two-stage programming is to minimize or maximize the sum of the expected cost or revenue of the first stage and the second stage. If the decision variables in the two-stage stochastic programming are required to be integer or binary, it is called a two-stage stochastic integer programming model.

This chapter is to introduce the solution algorithm used to solve the two-stage stochastic integer programming problems. Generally, two-stage stochastic integer programming (denoted by 2SSIP) problems can be formulated in the following way:

Liner Ship Fleet Planning
http://dx.doi.org/10.1016/B978-0-12-811502-2.00007-7

$$[2SSIP] \quad z_{2SSIP} = \min cx + \mathbb{E}_{\xi}[Q(x, \xi)] \tag{7.1}$$

$$\text{s.t.} \quad Ax = b, x \geq 0 \tag{7.2}$$

where $Q(x, \xi) = \min\{q(\xi)y | W(\xi)y \leq h(\xi) - B(\xi)x, y \geq 0\}$, and $\mathbb{E}_{\xi}$ denotes mathematical expectation with respect to ξ.

In which the vector x is called a *first-stage decision* and a vector y is called a *second-stage decision*. Vector ξ is to represent the random events. After the random events occurred and the information is realized, then a corrective action is then taken.

The two-stage stochastic programming model has three difficulties to be solved: First is the expected value function $\mathbb{E}_{\xi}[Q(x, \xi)]$, which does not have a closed form, so it is not easy to calculate its value. Secondly, the second-stage problem is formulated as an optimization model $Q(x, \xi)$ with uncertain parameters that are also not easy to be solved. Thirdly, the optimization model of the second-stage problem is nested in the optimization model of the first-stage problem, which also makes the two-stage stochastic programming model difficult to be solved.

In this chapter, we introduce the solution algorithm to solve the two-stage stochastic programming model. The solution algorithm is based on the dual decomposition scheme and Lagrangian relaxation. This solution algorithm has been investigated by Carøe and Schultz (1999). The following sections will discuss the solution algorithm in details. As the Lagrangian relaxation is used after the model is split by dual decomposition method, we firstly introduce the theoretical background of Lagrangian relaxation, then introduce the dual decomposition.

LAGRANGIAN RELAXATION

The Lagrangian relaxation method is an efficient method to deal with the discrete and combinatorial optimization problems that are hard to be solved, such as the knapsack problem, the traveling salesman problem, etc. This method was firstly proposed by Held and Karp (1970, 1971) to solve the traveling salesman problem. Following that, it is applied in the scheduling problems (Fisher, 1973) and the general integer programming problem (Fisher & Shapiro, 1974; Shapiro, 1971) as well. In 1974, Geoffrion named this method as "Lagrangian relaxation" (Geoffrion, 1974). Since then the success of Lagrangian relaxation method has attracted the attention of many researchers and has been applied in many optimization problems, especially in combinatorial optimization problems. Readers are referred to Fisher (2004) for the recent survey.

The Lagrangian relaxation method is to remove some constraints of the combinatorial optimization model in order to produce a Lagrangian problem that is easy to solve and whose optimal value is a lower bound (for minimization problem) on the optimal value of the original problem. The Lagrangian problem can thus be used in place of linear programming and offers a number of important advantages over that industry. The following sections will introduce the Lagrangian relaxation method in details.

Basic Constructions

Generally, a combinatorial optimization problem can be formulated as the following integer program

$$\begin{aligned} [\mathrm{P}] \quad & z = \min c^T x \\ \text{s.t.} \quad & Ax = b \\ & Dx \le e \\ & x \ge 0 \text{ and integral} \end{aligned} \tag{7.3}$$

where x is $n \times 1$, b is $m \times 1$, e is $k \times 1$ and all other matrices have conformable dimensions. If the problem [P] is relaxed with the integrality constraint on x and denoted by [LP], and the corresponding optimal value is denoted by z_{LP}, then it is easy to obtain the corresponding Lagrangian problem of [P]:

$$\begin{aligned} [\mathrm{LR}_u] \quad & z_D(u) = \min c^T x + u(Ax - b) \\ \text{s.t.} \quad & Dx \le e \\ & x \ge 0 \text{ and integral} \end{aligned} \tag{7.4}$$

where $u = (u_1, \ldots, u_m)$ is a vector of Lagrangian multipliers.

Theorem 7.1. The Lagrangian problem $[\mathrm{LR}_u]$ yields a lower bound to the original problem [P], i.e., $z_D(u) \le z$.

Proof. Let x^* be an optimal solution to the original problem [P], then for a given Lagrangian multiplier vector u, x^* is a feasible solution to the Lagrangian problem $[\mathrm{LR}_u]$, then we have $z_D(u) \le c^T x^* + u(Ax^* - b) = c^T x^* = z$. If $Ax = b$ is replaced by $Ax \le b$ in [P], then we can require the Lagrangian multiplier vector to be nonnegative; that is, $u \ge 0$, and the argument becomes $z_D(u) \le c^T x^* + u(Ax^* - b) \le z$. Similarly, for $Ax \ge b$ we require $u \ge 0$ for $z_D(u) \le z$ to hold. The theorem has been proven.

It is noted that the Lagrangian problem would yield an upper bound to the original problem for maximization problem.□

Subgradient Method to Determine Lagrangain Multiplier

It has been proved that the Lagrangian problem yields a lower bound to the original problem for a given Lagrangian multiplier vector u; consequently, it is obvious that we hope to find the best u in order to get the tightest lower bound. The best choice for u would be an optimal solution to the following problem, called the dual problem:

$$[\mathrm{D}] \quad z_D = \max u z_D(u) \tag{7.5}$$

There are many different methods to solve the dual problem [D]. Here, we introduce the subgradient method proposed by Fisher (1981), because this method has been applied to many problems successfully and has been proved to be an efficient and effective method.

An m-vector γ is called a subgradient of $z_D(u)$ at $\bar{u}$ if it satisfies

$$z_D(u) \le z_D(\bar{u}) + \gamma(u - \bar{u}), \quad \text{for all } u \tag{7.6}$$

The subgradient method is a brazen adaptation of the gradient method, in which gradients are replaced by subgradients. Given an initial value u^0, we can use the following rule to generate a sequence $\{u^k\}_1^\infty$

$$u^{k+1} = u^k + t_k\left(Ax^k - b\right) \tag{7.7}$$

where x^k is an optimal solution to the Lagrangian problem $[\mathrm{LR}_{u^k}]$ and t_k is a positive scalar step size in the kth iteration. The computational performance and theoretical convergence properties of the subgradient method are discussed in Held, Wolfe, and Crowder (1974) and Goffin (1977). Based on their studies, we have $z_D(u^k) \to z_D$ if $t_k \to 0$ and $\sum_{i=0}^{k} t_i \to \infty$. Therefore the Harmonic sequence can be a typical step size in the subgradient method:

$$t_k = \frac{1}{k} \tag{7.8}$$

Without generality, we can let the initial value $u^0 = 0$. The method is usually terminated upon reaching an arbitrary iteration limit, such as $\left|\left[z_D(u^{k+1}) - z_D(u^k)\right]/z_D(u^k)\right| < \delta$, where δ is a given threshold.

DUAL DECOMPOSITION

As mentioned above, this chapter introduces the solution algorithm of the dual decomposition method based on Lagrangian relaxation to solve the two-stage stochastic programming model. We have pointed out that there are three key difficulties in solving the two-stage stochastic programming models. The first difficulty is that the expected value function in the two-stage stochastic programming model is without a closed form and is therefore quite difficult to evaluate. So we first introduce the scenario approximation method to deal with the expected value function.

Scenarios

In Chapter 6, we introduce the method to deal with the chance-constrained programming model, in which the probability function in the chance constrains is replaced by the sample average approximation. The rationale behind the replacement is that the sample average can be used to approximate the probability based on the law of large numbers; it has been proved that the approximation can be convergent to the original problem. In the two-stage stochastic programming model, we can still employ this method to approximate the expected value function.

A realization of the random vector $(q(\xi), W(\xi), h(\xi), B(\xi))$ is defined as a scenario of an elementary event ξ. Typically, the distribution of ξ is multivariate, and we assume that we only have a finite number of r for scenarios of ξ for simplicity. The notation (q^j, W^j, h^j, B^j) is used to denote the jth scenario, which has a probability $p^j, j = 1, \ldots, r$. Therefore the expected value function in the two-stage stochastic programming model can be approximated as follows:

$$\mathbb{E}_\xi[Q(x, \xi)] \approx \sum_{j=1}^{r} p^j q^j y^j \tag{7.9}$$

It has to be noted that the problem (7.1) is equivalent to a mixed-integer programming problem with a large dimension and dual block-angularity, if the random vector ξ has a distribution with finite probability space. Define for $j = 1, \ldots, r$ the sets

$$S^j := \left\{ \left(x, y^j\right) : Ax \le b, x \in X;\ W^j y^j \le h^j - B^j x, y^j \in Y \right\} \tag{7.10}$$

Then the deterministic equivalent can be written as

$$z_{2SSIP}^{J} = \min\left\{cx + \sum_{j=1}^{r} p^{j}q^{j}y^{j} : (x, y^{i}) \in S^{j}, j = 1, \ldots, r\right\} \tag{7.11}$$

Decomposition

If we investigate the problem (7.11), we can find that the second term in its objective function is a summation operation of subproblems $p^{j}q^{j}y^{j}$ corresponding to each scenario j, while the first term in its objective function is cx corresponding to all scenarios. Such a structure with blocks makes us to split the summation for all scenarios into single piece for each scenario by using a decomposition approach; each piece will correspond to scenario sub-problems. Obviously, the idea for decomposing the summation operation based on the scenario is to introduce the copies $x^{1}, \ldots, x^{r}$ of the first-stage decision variable x, then rewrite Eq. (7.11) in the form

$$z_{2SSIP}^{J} = \min\left\{\sum_{j=1}^{r} p^{j}\left(cx^{j} + q^{j}y^{j}\right) : (x, y^{i}) \in S^{j},\ j = 1, \ldots, r \text{ and } x^{1} = \ldots = x^{r}\right\} \tag{7.12}$$

Here the additional constraint or condition that $x^{1} = \ldots = x^{r}$ is called nonanticipativity constraint, which indicates that the first-stage decision should not depend on the scenario that will prevail in the second stage. Such a requirement is necessary because the first-stage decision variable is here-and-now, which is to be determined before the second-stage decision variables, which is wait-and-see. The second-stage decision variables should be respective to the realization of the random vectors.

We can rewrite it as the equivalent form

$$\begin{aligned} x^{1} &= x^{2} \\ x^{2} &= x^{3} \\ &\vdots \\ x^{r+1} &= x^{r} \end{aligned} \tag{7.13}$$

It also can be rewritten as the matrix form

$$\sum_{j=1}^{r} H^{j}x^{j} = 0 \tag{7.14}$$

where H is a suitable matrix.

Lagrangian Dual

Now we can use the Lagrangian relaxation method presented above to move the nonanticipativity constraint (7.14) into the objective function in (7.12); then the Lagrangian relaxation to the original problem (7.12) would be as follows:

$$[\text{LR}-2\text{SSIP}] \quad z_{\text{LR}-2\text{SSIP}}(u) = \min \sum_{j=1}^{r} p^j \left(cx^j + q^j y^j\right) + uH^j x^j \tag{7.15}$$

where u is a vector of Lagrangian multipliers. The Lagrangian relaxation problem (7.15) is a convex, nonsmooth program that can be split into separate subproblems for each scenario; therefore, the following equation holds:

$$z_{\text{LR}-2\text{SSIP}}(u) = \sum_{j=1}^{r} z^j_{\text{LR}-2\text{SSIP}}(u), \tag{7.16}$$

where $z^j_{\text{LR}-2\text{SSIP}}(u) = \min p^j\left(cx^j + q^j y^j\right) + uH^j x^j$ for $j = 1, \ldots, r$. For the jth subproblem under the scenario j with a given Lagrangian multiplier vector u, the optimization model $z^j_{\text{LR}-2\text{SSIP}}(u)$ can be solved by an efficient optimization solver, as it is a manageable subproblem.

Obviously the LR-2SSIP model yields a lower bound to the 2SSIP based on Theorem 7.1. Furthermore, in order to find a tightest lower bound of the problem (7.15), we can solve the problem (7.17), which is called Lagrangian dual of problem (7.15). Then we have

$$[\text{LD}-2\text{SSIP}] \quad z_{\text{LD}-2\text{SSIP}} = \max_{u} z_{\text{LR}-2\text{SSIP}}(u) \tag{7.17}$$

The subgradient method is then used. Here the subgradient is $\sum_{j=1}^{r} H^j x^j$, where (x^j, y^j) is the optimal solution to the jth scenario subproblem. Based on the subgradient method introduced above, the Lagrangian multiplier can be updated by the following rule

$$u^{k+1} = u^k + t_k \sum_{j=1}^{r} H^j x^j \tag{7.18}$$

The step size and stop criterion can be as the same as described above.

SUMMARY

This chapter presents the solution algorithm based on a dual decomposition Lagrangian relaxation to solve the two-stage stochastic integer programming model. We firstly use the scenario average to approximate the expected value function. Secondly, we use the dual decomposition method to split the complicated summation operation of optimization resulting from the sample average approximation into single manageable pieces, in which the first-stage decision variables are copied a number of times to correspond to the number of scenarios in the second-stage. We then employ the Lagrangian relaxation method to deal with the nonanticipativity constraint, which is to keep the first-stage decision variables independent of the realization of scenarios. The solution algorithm is a combination of a dual decomposition and Lagrangian relaxation. Though it is complicated, it is an effective method to solve the two-stage stochastic integer programming model. The next part of this book will introduce four cases to show the applicability of stochastic models and proposed solution algorithms.

REFERENCES

Carøe, C. C., & Schultz, R. (1999). Dual decomposition in stochastic integer programming. *Operations Research Letters*, *24*, 37–45.

Fisher, M. L. (1973). Optimization solution of scheduling problems using Lagrange multipliers: Part I. *Operations Research*, *21*, 1114–1127.

Fisher, M. L. (1981). The Lagrangian relaxation method for solving integer programming problem. *Management Science*, *27*(1), 1–18.

Fisher, M. L. (2004). Comments on "The Lagrangian relaxation method for solving integer programming problems". *Management Science*, *50*(12), 1872–1874.

Fisher, M. L., & Shapiro, J. F. (1974). Constructive duality in integer programming. *SIAM Journal of Applied Mathematics*, *27*, 31–52.

Geoffrion, A. M. (1974). Lagrangian relaxation and its uses in integer programming. *Mathematical Programming Studies*, *2*, 82–114.

Goffin, J. L. (1977). On the convergence rates of subgradient optimization methods. *Mathematical Programming*, *13*, 329–347.

Held, M., & Karp, R. M. (1970). The traveling salesman problem and minimum spanning trees. *Operations Research*, *18*, 1138–1162.

Held, M., & Karp, R. M. (1971). The traveling salesman problem and minimum spanning trees: Part II. *Mathematical Programming*, *1*, 6–52.

Held, M., Wolfe, P., & Crowder, H. D. (1974). Validation of subgradient optimization. *Mathematical Programming*, *6*, 62–88.

Shapiro, J. F. (1971). Generalized Lagrange multipliers in integer programming. *Operations Research*, *19*, 68–76.

PART FOUR

Case Studies

CHAPTER EIGHT

Liner Ship Fleet Planning Problem With an Individual Chance-Constrained Service Level

Contents

INTRODUCTION

This chapter deals with the short-term LSFP problem encountered by a liner container shipping company. The liner container shipping company or liner operator usually operates a fleet of heterogeneous ships on its service routes at regular intervals in order to pick up and deliver containers for shippers. In order to seize market share in an intensely competitive container shipping market, the liner container shipping company is constantly searching for models and solution procedures to build a decision support system that helps to create cost-effective plans to operate its liner ship fleet.

http://dx.doi.org/10.1016/B978-0-12-811502-2.00008-9

In addition, the number of containers transported by a liner container shipping company between two ports often varies season by season (i.e., every 3 months) in practice. For example, the container volume from Asia to Europe usually increases dramatically in the fourth quarter of a year due to Christmas Day. To cope with the varying port-to-port container shipment demand, a liner container shipping company has to alter its service routes and redeploy ships according to the estimated container shipment for the next season. In other words, the company's strategic asset management department needs to make a suitable fleet plan for a short-term (3–6 months) planning horizon, which involves considering how to effectively use the ships in its fleet in order to provide efficient shipping services and save on costs. The decisions include the determinations of fleet size (number of ships), mix (ship types), and deployment (ship-to-route assignment). For the sake of presentation, this tactical-level decision is referred to as the short-term liner ship fleet planning (LSFP) problem. The aim is to optimize fleet design and deployment over a short-term planning horizon. The fleet design identifies the types and numbers of ships required, and the fleet deployment covers how the fleet is assigned and operated to transport containers.

Container shipment demand of a port pair on a liner ship route operated by the liner container shipping company is an input of the short-term LSFP problem. The decisions of fleet size, mix, and fleet deployment involved in this problem are made prior to knowing the exact market demand. Liner shipping is usually based on a fixed schedule, which is generally published up to 6 months into the future. This means the LSFP is made depending on the forecasted container shipment demand. The container shipment demand is usually estimated by some shipment demand forecasting methods. Compared with the actual port-to-port container shipment demand, the forecasted shipment demand is inevitably biased because it is usually affected by some unpredictable and uncontrollable factors. Hence the container shipment demand is of high uncertainty in practice. This chapter thus focuses on model development for the short-term LSFP problem by taking into consideration the uncertainty of container shipment demand.

It should be pointed out that the existing studies on fleet size and mix problems (i.e., Dantzig & Fulkerson, 1954; Fagerholt, 1999; Fagerholt & Lindstad, 2000; Lane, Heaver, & Uyeno, 1987; Sambracos, Paravantis, Tarantilis, & Kiranoudis, 2004), fleet deployment problems (i.e., Benford, 1981; Bradley, Hax, & Magnanti, 1977; Laderman, Gleiberman, & Egan, 1966; Mourão, Pato, & Paixão, 2001; Papadakis & Perakis, 1989;

Perakis, 1985; Perakis & Papadakis, 1987a, 1987b); fleet planning problems (i.e., Everett, Hax, Lewinson, & Nudds, 1972; Gelareh & Meng, 2010; Perakis & Jarammillo, 1991; Powell & Perakis, 1997) assume that the container shipment demand between two ports are deterministic. As discussed previously, the container shipment demand has high uncertainty in practice. This uncertainty can be formulated as the stochastic container shipment demand represented by a random variable. Having assumed the stochastic container shipment demand, the aforementioned three categories of the decision problems should be reformulated. The existing linear or integer models reviewed above for the deterministic LSFP problems are not applicable for the proposed problem. Therefore a new stochastic programming model for uncertain container shipment demand is needed to formulate this problem.

In this chapter the container shipment demand between any two ports on each liner ship route is assumed to follow a normal distribution; the probability (chance) that shipping capacity of a LSFP scenario cannot meet the demand does exist. In other words the liner container shipping company fails to make the service for its customers with this probability. To maintain a certain level of service, the company must control this probability (or chance) within a given level. This is called the confidence parameter; the level of service is referred to as chance constraint hereafter. Therefore a chance-constrained programming (CCP) model is proposed for the short-term LSFP problem with container shipment demand uncertainty. The objective of this CCP model is to minimize the total operating cost of its fleet subject to a certain level of service.

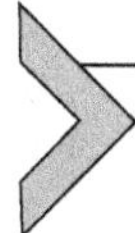

PROBLEM DESCRIPTION, ASSUMPTIONS, AND NOTATIONS

Code of Port Sequence

Consider a liner container shipping company that provides a liner shipping service on a predetermined liner ship route network for shippers within a short-term planning horizon (e.g., 3–6 months). Let $\mathcal{P}=\{1,\ldots,p,\ldots,P\}$ and $\mathcal{R}=\{1,\ldots,r,\ldots,R\}$ denote the set of ports and the set of liner ship routes in the liner ship route network, respectively. The indices p and r represent a particular port and liner ship route, respectively. Additionally, we define $\mathcal{P}_r=\{p_r^1,\ldots,p_r^i,\ldots p_r^{m_r}\}$ as the set of ports called at the liner ship route $r\in\mathcal{R}$, characterized by $\mathcal{P}=\bigcup_{r\in\mathcal{R}}\mathcal{P}_r$, where m_r is the number of ports in the itinerary. Each liner ship route $r \in \mathcal{R}$ is defined as a sequence of ports called at by ships, which can be expressed by the port calling sequence (or itinerary):

$$p_r^1 \rightarrow p_r^2 \rightarrow \cdots \rightarrow p_r^{m_r} \rightarrow p_r^1 \tag{8.1}$$

Eq. (8.1) describes the unique characteristic of a liner ship route: a loop with a given port calling order. Note that the ports on a liner ship route may not all be distinct. For example, Fig. 8.1 depicts a liner ship route between the port of Pusan and the port of Singapore. A ship deployed on this liner ship route first calls at Pusan (PS), followed by Shanghai (SH), Yantian (YT), Hong Kong (HK), Singapore (SG), YT, and finally back to PS. According to the route coding scheme shown in Eq. (8.1), this can be expressed by the port calling sequence:

$$\begin{aligned} &p_r^1(\text{PS}) \rightarrow p_r^2(\text{SH}) \rightarrow p_r^3(\text{YT}) \rightarrow p_r^4(\text{HK}) \rightarrow p_r^5(\text{SG}) \\ &\rightarrow p_r^6(\text{YT}) \rightarrow p_r^1(\text{PS}) \end{aligned} \tag{8.2}$$

Fig. 8.1 also shows that the port calling sequence for the forward direction from Pusan to Singapore is not identical to that for the backward direction from Singapore to Pusan.

Container Shipment Flow

To formulate the feature that the first port and last port on a liner ship route are the same, we introduce a generalized mod operator as follows:

$$i \,\underline{\text{mod}}\, m_r = \begin{cases} i \bmod m_r, & i \neq m_r \\ m_r, & i = m_r \end{cases} \tag{8.3}$$

For the sake of presentation, the distance between two consecutive ports p_r^i and $p_r^{(i+1)\underline{\text{mod}}\,m_r}$ on the liner ship route $r \in \mathcal{R}$ is referred to as a leg i $(i=1,2,\ldots,m_r)$, denoted by the pair of ordered ports $<p_r^i, p_r^{(i+1)\underline{\text{mod}}\,m_r}>$. The liner ship route shown in Eq. (8.2) thus has six legs—1:

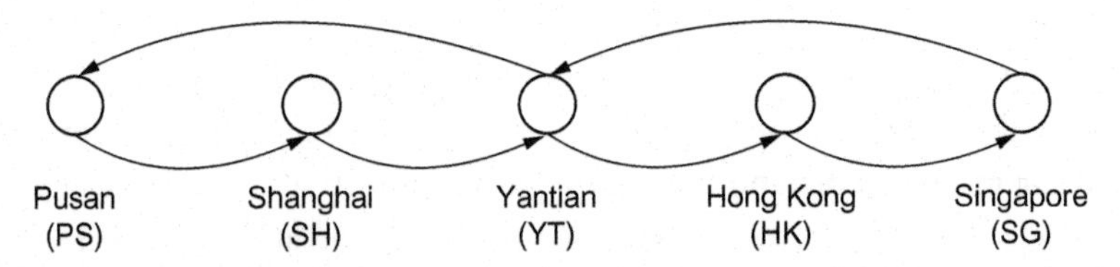

Fig. 8.1 A liner ship route.

$<p_r^1(\mathrm{PS}), p_r^2(\mathrm{SH})>$, 2: $<p_r^2(\mathrm{SH}), p_r^3(\mathrm{YT})>$, 3: $<p_r^3(\mathrm{YT}), p_r^4(\mathrm{HK})>$, 4: $<p_r^4(\mathrm{HK}), p_r^5(\mathrm{SG})>$, 5: $<p_r^5(\mathrm{SG}), p_r^6(\mathrm{YT})>$, and 6: $<p_r^6(\mathrm{YT}), p_r^1(\mathrm{PS})>$.

The port calling sequence shown in Eq. (8.1) has a limited number of combinations of port pairs, which may have container shipment demand on the liner shipping service route $r \in \mathcal{R}$. These pairs of ports can be expressed by the set

$$\mathcal{M}_r = \left\{ \left(p_r^i, p_r^j\right) \middle| i,\ j = 1, 2, \ldots, m_r;\ p_r^i \neq p_r^j \right\} \tag{8.4}$$

An incidence parameter $\rho_l^{(p_r^i, p_r^j)}$ is defined to indicate how the containers are transported from port p_r^i to port p_r^j, namely the itinerary of transporting containers of port pairs (p_r^i, p_r^j). It equals 1 if leg l is sailed by ships transporting containers from port p_r^i to port p_r^j; otherwise, it is 0 $(r \in \mathcal{R})$. The incidence parameter thus reflects the relationship between the itinerary for transporting containers from port p_r^i to port p_r^j and the legs l in the liner shipping service route. We use the above example to illustrate this. Containers being transported between the port pair (p_r^2, p_r^1), namely from Shanghai to Pusan, have to be loaded at port p_r^2 (Shanghai) and then carried by ships along legs 2, 3, 4, 5, and 6, before finally being unloaded at port p_r^1 (Pusan). Therefore $\rho_1^{(p_r^2, p_r^1)} = 0$ and $\rho_l^{(p_r^2, p_r^1)} = 1\ (l = 2, 3, 4, 5, 6)$.

When a ship sails along leg l of route r, it carries containers, including those new containers loaded at port p_r^l and also those loaded at previous ports that have remained on the ship, which is referred to as container shipment flow on leg l. Continuing with the above example, when a ship sails on leg 6, it carries containers corresponding to eight port pairs: (p_r^2, p_r^1), (p_r^3, p_r^1), (p_r^4, p_r^1), (p_r^5, p_r^1), (p_r^3, p_r^2), (p_r^4, p_r^2), (p_r^5, p_r^2), and (p_r^5, p_r^4), of which the containers being transported between the port pair (p_r^3, p_r^1) and between the pair (p_r^3, p_r^2) were newly loaded at port p_r^3, and the containers for the other six port pairs were loaded at previous ports.

Liner Ship Fleet Planning

Let $\mathcal{K} = \{1, \ldots, k, \ldots, K\}$ be the set of ship types available to the liner container shipping company, where the index k denotes a particular type of ships. The container capacity in terms of twenty-foot-equivalent unit (TEU) of a particular ship type k is denoted by V_k. The liner container shipping company has to determine the number of ships of type $k \in \mathcal{K}$ in its ship fleet and deploy them on each liner ship route $r \in \mathcal{R}$ to pick up and deliver containers for shippers at a regular schedule on each route.

In the short-term LSFP problem, the liner container shipping company not only uses its own ships to deliver containers, but it also charters ships from other liner shipping companies. Generally, there are three types of chartering ships: bareboat charter, voyage charter, and time charter. Bareboat charter is the simplest way in which the charterer manages the ship and pays all costs except the capital repayment, tax, and depreciation. In other words, the owner does not bear any cost except collecting the rent from the charterer. In order to simplify the problem, bareboat charter is the one only adopted in this thesis. The chartering rate of a ship of type k in the planning horizon is denoted by c_k^{IN} (\$/ship). Besides paying the chartering rate to the ship owner, the ship charterer takes on other charges of operating the chartered ship, such as routine maintenance cost, insurance, etc. Therefore it is rational to assume that

$$c_k^{\mathrm{OUT}} < c_k^{\mathrm{IN}} \tag{8.5}$$

where c_k^{OUT} denotes the rate of a ship of type k chartered out (\$/ship).

Let N_k^{MAX} and NCI_k^{MAX} denote the number of available ships of type k owned and chartered by the liner container shipping company, respectively. Given these candidate ships, the liner container shipping company chooses some ships to form a liner ship fleet; namely, a ship fleet design plan comprising the mix and size of the ship fleet, and then assigns the ships in the fleet to those ship routes. The objective is to make an efficient joint ship fleet design and ship fleet deployment plan in order to maximize the expected value of the total profits subject to some constraints. It is noted that a regular shipping service is required to be maintained on each ship route because liner shipping is characterized by providing regular shipping service in contrast to tramp and industrial shipping (Christiansen, Fagerholt, & Ronen, 2004). In practice, most liner shipping companies generally provide a weekly shipping service on a ship route.

Container Shipment Demand Uncertainty

In practice the container shipment demand over the short-term planning horizon T (3–6 months) is estimated by using some demand forecast methods based on historical data. However, the estimated container shipment demand is biased, and it thus causes uncertainty of the container shipment demand used for a short-term planning decision. Uncertainty of container shipment demand comes also from transactions between shippers and the liner container shipping company, briefed as follows: A shipper firstly needs to book space from the shipping company according to the ship schedule and itinerary launched by the company in order to deliver its containers through a shipping agent by filling in a shipping application (S/A). If the S/A is accepted the shipper will receive a shipping order (S/O) from the shipping company to load its containers on a ship operated by the shipping company. Then the carrier (i.e., the liner container shipping company) will offer a mate's receipt (M/R) to the shipper to show its containers are loaded on the ship. The shipper bears the M/R to exchange the bill of lading (B/L) and posts it to the consignee. The shipping agent at the discharge port informs the consignee to retrieve the containers when they arrive. After the payment of all fees, the consignee uses B/L to exchange the delivery order (D/O) and takes delivery of the goods. However, the shipper is allowed to cancel the transaction or contract signed with the shipping company in advance. The cancelation as an uncontrollable factor brings uncertainty of the estimated container shipment demand.

The distribution-based approach is a typical method to characterize the parameter uncertainty issue. It is usually used to describe the issue with exact concept or essence, but whether that happens depends on some random factors. For example, when flipping a coin, the concept of "face of coin" is exact, but the occurrence is related to some unpredictably or uncontrollably uncertain factors. The container shipment demand uncertainty considered in this chapter has an exact concept, but it is related to some random factors; therefore the distribution-based approach is reasonable to be employed to formulate uncertainty of container shipment demand. Following the distribution-based uncertainty characterization approach, the container shipment demand is assumed normally distributed with given mean and standard deviation. The rationale of assuming the normal distribution is that the normal distribution has been established to be one of the suitable probability distribution to describe the demand uncertainty by Brown (1959). Without a loss of generality, these normal random container shipment demands are assumed to be independent.

Problem Statement

As already mentioned, the container shipment demand between any two ports on each liner ship route is assumed to follow a normal distribution. However, this assumption may lead to another problem: As the demands are uncertain, one can hardly find any decision which would definitely exclude a later constraint violation caused by unexpected random effects. In other words, once the decisions in LSFP problem are determined, the fleet of ships may be unable to fully meet the pickups and deliveries requirement for its customers, even though the expected demands along the route do not exceed the fleet capacity. Once such a case happens, it implies losing money for this liner container shipping company. As it is hardly unavoidable, the liner container shipping company has to hope that it happens at the lowest possibility.

In order to reduce the possibility of the occurrence that the liner container shipping company cannot satisfy the customers' demand, such a constraint violation can often be balanced afterwards by some compensating decisions, which are considered as a penalization for constraint violation. However, the compensation cannot be modeled by cost in this chapter because the container shipment demand is not realized. In such circumstances, we would rather insist on decisions guaranteeing feasibility "as much as possible." This loose term refers once more to the fact that constraint violation can almost never be avoided because of unexpected extreme events. On the other hand, when knowing or approximating the distribution of the random parameter, it makes sense to call decisions feasible whenever they are feasible with high probability; that is, only a low percentage of realizations of the random parameter leads to a constraint violation under this fixed decision. Therefore, we formulate the constraint that the liner container shipping company should satisfy the customers' demand as a probabilistic form in this chapter, which is called chance constraint. The probability of the constraint violation is called a confidence parameter in this chance constraint. It indicates that if the liner container shipping company makes a decision which satisfies the chance constraint, the event that the customers' demand cannot be met will occur at most with this probability. For those cargoes which cannot be shipped by carriers due to limited shipping capacity, we regarded they are lost.

Therefore the short-term LSFP problem with container shipment demand uncertainty aims to determine the best decision variables to minimize the total operating cost while maintaining the chance constraints. It is formulated as a CCP model.

MODEL DEVELOPMENT

Before the development of mathematical programming model for the short-term LSFP problem with container shipment demand uncertainty, we firstly introduce the decision variables shown as follows:

n_{kr}^{OWN}, number of owned ships of type k $(k \in \mathcal{K})$ assigned on route r $(r \in \mathcal{R})$
n_{kr}^{IN}, number of chartered in ships of type k $(k \in \mathcal{K})$ assigned on route r $(r \in \mathcal{R})$
x_{kr}, number of voyages of ships of type k $(k \in \mathcal{K})$ on route r $(r \in \mathcal{R})$

Chance Constraints for Individual Service Level

Let $\xi^{(p_r^i, p_r^j)}$ be the random variable representing the container shipment demand of a port pair $(p_r^i, p_r^j) \in \mathcal{M}_r$, while the container shipment flow on leg l $(l=1,\ldots,m_r)$ of route $r \in \mathcal{R}$, denoted by η_l^r, is given by:

$$\eta_l^r = \sum_{(p_r^i, p_r^j) \in \mathcal{M}_r} \rho_l^{(p_r^i, p_r^j)} \xi^{(p_r^i, p_r^j)}, \quad l=1,\ldots,m_r; \ \forall r \in \mathcal{R} \tag{8.6}$$

As the container shipment demand of any port pair, $\xi^{(p_r^i, p_r^j)}$ $\left((p_r^i, p_r^j) \in \mathcal{M}_r, r \in \mathcal{R}\right)$, is assumed following an inter-independent normal distribution with a mean value denoted by $\mu^{(p_r^i, p_r^j)}$ and a variance denoted by $\sigma^{(p_r^i, p_r^j)^2}$, namely $\xi^{(p_r^i, p_r^j)} \sim N\left(\mu^{(p_r^i, p_r^j)}, \sigma^{(p_r^i, p_r^j)^2}\right)$, according to the probability theory, η_l^r also follows a normal distribution, so we then have:

$$\eta_l^r \sim N\left(\sum_{(p_r^i, p_r^j) \in \mathcal{M}_r} \rho_l^{(p_r^i, p_r^j)} \mu^{(p_r^i, p_r^j)}, \sum_{(p_r^i, p_r^j) \in \mathcal{M}_r} \rho_l^{(p_r^i, p_r^j)} \sigma^{(p_r^i, p_r^j)^2}\right), \tag{8.7}$$
$$l=1,\cdots,m_r; \ \forall r \in \mathcal{R}$$

Let α_r denote the confidence parameter on route r; therefore the constraints that the transportation capacity of containerships operated on this route is not less than each container shipment flow on leg l with a least probability of $1-\alpha_r$ can be formulated as the following chance constraints:

$$\Pr\left(\sum_{k \in \mathcal{K}} x_{kr} V_k \geq \eta_l^r\right) \geq 1-\alpha_r, \quad l=1,\ldots,m_r; \ \forall r \in \mathcal{R} \tag{8.8}$$

CCP Model

The proposed short-term LSFP problem with uncertain container shipment demand can be formulated as the CCP model:

$$\text{[CCP]} \quad \min\ C = \sum_{r\in\mathcal{R}} \sum_{k\in\mathcal{K}} \left(c_{kr} x_{kr} + n_{kr}^{\mathrm{IN}} c_k^{\mathrm{IN}} \right) \tag{8.9}$$

subject to

$$\Pr\left(\sum_{k\in\mathcal{K}} x_{kr} V_k \geq \eta_l^r \right) \geq 1 - \alpha_r, \quad l = 1, \ldots, m_r; \ \forall r \in \mathcal{R} \tag{8.10}$$

$$x_{kr} \leq \left(n_{kr}^{\mathrm{OWN}} + n_{kr}^{\mathrm{IN}} \right) \times \left\lfloor \frac{T}{t_{kr}} \right\rfloor, \forall r \in \mathcal{R}, k \in \mathcal{K} \tag{8.11}$$

$$\sum_{k\in\mathcal{K}} x_{kr} \geq N_r, \forall r \in \mathcal{R} \tag{8.12}$$

$$\sum_{r\in\mathcal{R}} n_{kr}^{\mathrm{OWN}} \leq N_k^{\mathrm{MAX}}, \forall k \in \mathcal{K} \tag{8.13}$$

$$\sum_{r\in\mathcal{R}} n_{kr}^{\mathrm{IN}} \leq NCI_k^{\mathrm{MAX}}, \forall k \in \mathcal{K} \tag{8.14}$$

$$n_{kr}^{\mathrm{OWN}}, n_{kr}^{\mathrm{IN}}, x_{kr} \in \mathbb{Z}^+ \cup \{0\}, \forall k \in \mathcal{K}, r \in \mathcal{R} \tag{8.15}$$

where c_{kr} denotes the operating cost of ships of type k on route r per voyage ($/voyage). It includes the fuel cost, daily running cost, port charge, and canal fee (if any). The chartering rate of a ship of type k in the planning horizon is denoted by c_k^{IN} ($/ship). t_{kr} is the voyage time of a ship of type k on a route r (days), T is the length of the short-term planning horizon (3–6 months), N_r is the minimal number of voyages required on route r during the planning horizon in order to maintain a given liner shipping service frequency, and V_k denotes the capacity of a ship of type k referring to the number of containers it can be loaded.

Eq. (8.9) is the objective function of the CCP model. The first term in the bracket presents the shipping cost, and the second term is the cost of chartering in containerships of type k in the short-term planning horizon. Constraints (8.10) are the chance constraints that show the liner container shipping company can satisfy the customers' demand at least with a probability of $1 - \alpha_r$. Constraints (8.11) compute the maximal number of voyage that ships of type k can complete on route r, where $\lfloor a \rfloor$ denotes the maximum integer not greater than a. The constraints (8.12) guarantee the number of voyages required on ship route r in order to maintain the given liner shipping frequency. For example, if a weekly shipping service is required on ship route r during a planning horizon of 6 months, then $N_r = 26$. Constraints (8.13) and (8.14) ensure the number containerships of their own plus those chartered in should not exceed its corresponding maximum available containerships, respectively. Constraint (8.15) requires that all variables are nonnegative integers.

It is not difficult to find that constraints (8.10) can be respectively rewritten as follows:

$$\sum_{k\in\mathcal{K}} x_{kr} V_k \geq \Phi^{-1}(1-\alpha_r)\sqrt{\sum_{(p_r^i, p_r^j)\in\mathcal{M}_r} \rho_l^{(p_r^i, p_r^j)} \sigma^{(p_r^i, p_r^j)^2}} + \sum_{(p_r^i, p_r^j)\in\mathcal{M}_r} \rho_l^{(p_r^i, p_r^j)} \mu^{(p_r^i, p_r^j)}, \quad l=1,\ldots,m_r; \forall r\in\mathcal{R}, \tag{8.16}$$

where $\Phi^{-1}(1-\alpha_r)$ is the inverse cumulative probability of $1-\alpha_r$. Eq. (8.16) imply that constraints (8.10) have the equivalent linear function expressions. Objective function shown by Eq. (8.9) and the other constraints (8.11)–(8.15) are all linear functions with respect to the decision variables. Therefore the CCP model is an integer linear programming model. As the CCP model is an integer linear programming model, it can be thus solved by any optimization solver, such as CPLEX. CPLEX actually employs the branch-and-cut algorithm for solving an integer linear programming problem.

The proposed CCP model involves two blocks of costs: those for shipping and those for chartering in. The shipping costs include fuel cost, daily running cost, port charge, and canal fee. The rationale behind port charges is that port authorities levy various fees against ships and/or containers for the use of the facilities and services provided by them; and the main canal dues payable are for transiting the Suez and Panama canals. As for chartering in ships, it is commonly adopted by liner shipping companies in practice. For example, APM-Maersk, the largest maritime container shipping operator in the world, operates totally 524 ships, in which it owns 184 ships and charters 340 ships in 2007. Therefore chartering-in costs are also included.

NUMERICAL EXAMPLE

In this section, we use a numerical example to assess the CCP model. Then taking this numerical example as a benchmark pattern, we investigate the impact of container shipment demand and the confidence parameter on the optimal decisions made in the proposed short-term LSFP problem.

Example Design

In the numerical example, we assume that a liner container shipping company intends to make a six-month fleet plan. In order to make the example close to a realistic case, we design a liner ship route network consisting of eight routes operated by a liner container shipping company, OOCL in Hong Kong. The liner shipping topology involves a total of 36 calling ports and 390 O-D pairs. The ports called on each liner ship route and their digital number codes are shown in Table 8.1. Table 8.2 gives the distance of each leg in each liner ship route. The numbers, sizes, market prices, daily operating cost, and the design speed of each ship type are listed in Table 8.3. It is noted that the daily operating cost of each ship type is estimated by using the following regression equation (Shintani, Imai, Nishimura, & Papadimitriou, 2007), as the exact data is unavailable:

$$\text{daily operating cost} = 6.54 \times \text{ship size(TEU)} + 1422.5 \tag{8.17}$$

Although this example is hypothetical, it is close to a "realistic" case. This is because some data of the numerical example are extracted from a real liner shipping company-OOCL; for example, calling ports in a liner ship route, types of ships and their sizes, sailing speeds, and so on. However, some data is still unavailable, including miscellaneous shipping costs and container

Table 8.1 Port Calling Sequence and Number Code for Each Route

Routes	Port Calling Sequence and Number Code
CCX	Los Angeles/Oakland/Pusan/Dalian/Xingang/Qingdao/Ningbo/Shanghai/Pusan/Los Angeles (1-2-3-4-5-6-7-8-9-1)
CPX	Shanghai/Ningbo/Shekou/Singapore/Karachi/Mundra/Penang/Port Kelang/Singapore/Hong Kong/Shanghai (1-2-3-4-5-6-7-8-9-10-1)
GIS	Singapore/Port Kelang/Nhava Sheva/Karachi/Jebel Ali/Bandar Abbas/Jebel Ali/Mundra/Cochin/Singapore (1-2-3-4-5-6-7-8-9-1)
IDX	Colombo/Tuticorin/Cochin/NhavaSheva/Mundra/Suez/Barcelona/New York/Norfolk/Charleston/Barcelona/Suez/Colombo (1-2-3-4-5-6-7-8-9-10-11-12-1)
NCE	New York/Norfolk/Savannah/Panama/Pusan/Dalian/Xingang/Qingdao/Ningbo/Shanghai/Panama/New York (1-2-3-4-5-6-7-8-9-10-11-1)
NZX	Singapore/Port Kelang/Brisbane/Auckland/Napier/Lyttelton/Wellington/Brisbane/Singapore (1-2-3-4-5-6-7-8-1)
SCE	New York/Norfolk/Savannah/Panama/Kaohsiung/Shekou/Hong Kong/Panama/New York (1-2-3-4-5-6-7-8-1)
UKX	Southampton/Hull/Grangemouth/Southampton (1-2-3-1)

Table 8.2 Distances of Each Leg in Each Liner Ship Route

Routes	Distance (Nautical Miles)
CCX	298-4985-543-158-356-339-60-492-5230
CPX	60-845-1460-2887-261-2510-172-210-1460-845
GIS	210-3097-261-711- 151-151-962-953-1853
IDX	140-161-158-186-2809-1673-3721-287-429-4124-1673-3394
NCE	287-505-982-13,831-543-158-356-339-60-13,565-1359
NZX	210-4050-1358-377-336-174-1448-3840
SCE	287-505-982-12949-342-33-12,788-1359
UKX	324-256-528

Table 8.3 Example Data

	Ship Types				
Item	**1**	**2**	**3**	**4**	**5**
Ship size (TEUs)	2808	3218	4500	5714	8063
Daily cost (10^3 $)	19.8	22.5	30.9	38.8	54.2
Design speed (knots)	21.0	22.0	24.2	24.6	25.2
Chartering-in rate (million $)	2	2.6	3.5	4.7	6.0
N_k^{MAX}	2	2	9	2	12
NCI_k^{MAX}	5	5	3	5	5

shipment demand between two ports on a liner ship route, as they are business confidential. These data are thus determined in a reasonable manner. As for the data of the container shipment demand, though OOCL provides the annual business report, the port-to-port container shipment demand on a liner ship route is not elaborated upon, so the data are hypothetical in this example. As the data of miscellaneous shipping costs and port-to-port container shipment demand are too many (more than one thousand), they are not listed for reasons of space.

CCP Model Assessment

Table 8.4 shows the confidence-parameter predetermined set on each route. With liner shipping services at level $1-\alpha$ (α is given in Table 8.4), the optimal solution of fleet size, mix, and deployment for this example is obtained by CPLEX Ver. 11 and shown in Table 8.5. It can be seen from Table 8.4 that the confidence parameters set on Route NCE and Route SCE are small, which indicates that a high level of service has to be maintained on these two routes. Therefore most ships are allocated to these two routes in order to maintain the high level of service.

Table 8.4 Confidence Parameters on Each Liner Ship Route

	Route							
	CCX	**CPX**	**GIS**	**IDX**	**NCE**	**NZX**	**SCE**	**UKX**
α	0.10	0.15	0.05	0.15	0.05	0.10	0.05	0.15

Table 8.5 Results to Benchmark Pattern

	Route							
Ship Type	**CCX**	**CPX**	**GIS**	**IDX**	**NCE**	**NZX**	**SCE**	**UKX**
Ship allocations								
1					1	2		1
2			2					
3	2	2				1	6	
4			1	2			2	
5				4	8			
Number of voyages								
1					8	17		26
2			16					
3	26	26				10	21	
4			10	14			8	
5				22	24			

Sensitivity Analysis

To study the impact of container shipment demand, 10 sets of container shipment demands are tested with the same confidence parameters in Table 8.4. These 10 sets of container shipment demands are generated by setting 60%, 70%, and up to 150% of the benchmark demand pattern. The trend of corresponding optimal objective function value with each of these 10 sets is shown in Fig. 8.2. This figure indicates that with the increase of the container shipment demand, more costs are taken to maintain the same level of service. The ratios of costs corresponding to other sets with the costs of benchmark demand pattern increase from 70% to 130% shown in Fig. 8.3. We take three sets of different confidence parameters as shown in Table 8.6 to analyze their impacts on the optimal fleet planning solution.

As the level of service equals $1 - \alpha$, it implies that a lower level of service corresponds to a larger value of α. Hence Set 1 in Table 8.6 indicates a low level of service, while Set 2 shows a medium level of service, and Set 3 suggests a high level of service. The optimal solutions corresponding to these

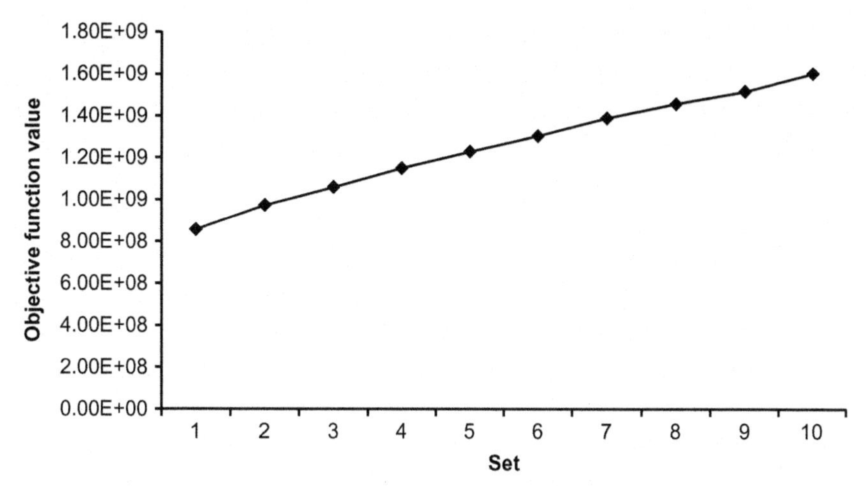

Fig. 8.2 Objective function value for different container shipment demand sets.

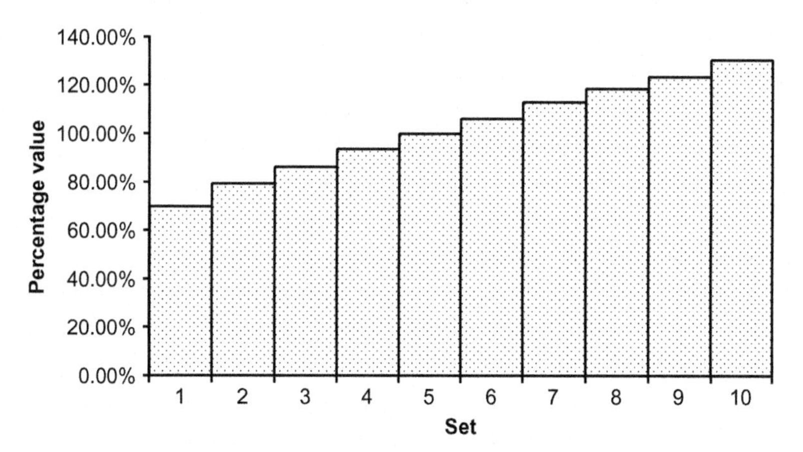

Fig. 8.3 Ratio of optimal objective function value for different sets with benchmark pattern.

Table 8.6 Three Sets of Confidence Parameters

	Route							
α	CCX	CPX	GIS	IDX	NCE	NZX	SCE	UKX
Set 1	0.20	0.15	0.20	0.10	0.15	0.20	0.15	0.15
Set 2	0.15	0.10	0.15	0.05	0.10	0.15	0.10	0.10
Set 3	0.10	0.05	0.10	0.05	0.05	0.10	0.05	0.05

three sets of confidence parameters are listed in Tables 8.7–8.9, respectively. These three tables imply that the confidence parameter has a significant impact on the optimal fleet size and deployment. It can be also found that more ships are needed and more costs must be taken in order to maintain a higher level of service.

Table 8.7 Results With Respect to Confidence Parameters in Set 1

	Route							
Ship Type	**CCX**	**CPX**	**GIS**	**IDX**	**NCE**	**NZX**	**SCE**	**UKX**
Ship allocations								
1					1	2		1
2			2					
3	2	2				1	6	
4			1	2			2	
5				4	8			
Number of voyages								
1					8	17		26
2			20					
3	26	26				10	21	
4			8	14			8	
5				22	25			
Cost (million $)	949.5924							

Table 8.8 Results With Respect to Confidence Parameters in Set 2

	Route							
Ship Type	**CCX**	**CPX**	**GIS**	**IDX**	**NCE**	**NZX**	**SCE**	**UKX**
Ship allocations								
1						2		1
2			2					
3	2	3			1	1	6	
4			1	2			2	
5				4	8			
Number of voyages								
1						17		26
2			20					
3	26	26			5	10	21	
4			7	14			8	
5				22	24			
Cost (million $)	1162.2142							

Table 8.9 Results With Respect to Confidence Parameters in Set 3

	Route							
Ship Type	**CCX**	**CPX**	**GIS**	**IDX**	**NCE**	**NZX**	**SCE**	**UKX**
Ship allocations								
1						2		1
2			2					
3	3	3			1	1	6	
4			1	2			2	
5				4	8			
Number of voyage								
1						17		26
2			24					
3	27	27			3	10	24	
4			3	14			6	
5				22	29			
Cost (million $)	1340.1157							

SUMMARY

This chapter makes an initiative to investigate the container shipment demand uncertainty issue rising from practice for the short-term liner fleet planning problems. Assuming that the container shipment demand of a port pair on each liner ship route follows a normal distribution, the probability (chance) does exist that the shipping capacity of a LSFP scenario cannot meet the demand. In other words, the liner container shipping company is failed to make the service for its customers with this probability. The level of service is proposed to represent the probability of satisfying the customers' requirement, and it can be formulated as a chance constraint. To maintain a certain level of service, the company must control this probability within a given level called a confidence parameter. We therefore develop a CCP model for the short-term LSFP problem with a container shipment demand uncertainty. The proposed model can be solved by many optimization solvers such as CPLEX because it is an integer linear programming model. A numerical example has been carried out for the model assessment and impact analysis of the confidence parameters and cargo shipment demand.

It can be seen that the service level in this chapter is for an individual route, not for the whole shipping network; see the chance constraints (8.10). In this case, the service level for the whole shipping network would be quite low because it is the product of the service level on each route. As for the service level on the whole shipping network, we will present the formulation and methodology in next chapter.

REFERENCES

Benford, H. (1981). A simple approach to fleet deployment. *Maritime Policy and Management, 8*(4), 223–228.
Bradley, S. P., Hax, A. C., & Magnanti, T. L. (1977). *Applied mathematical programming*. Reading, MA: Addison-Wesley [chapter 7].
Brown, R. G. (1959). *Statistical forcasting for inventory control*. New York, NY: McGrav-Hill.
Christiansen, M., Fagerholt, K., & Ronen, D. (2004). Ship routing and scheduling: Status and perspectives. *Transportation Science, 38*(1), 1–18.
Dantzig, G. B., & Fulkerson, D. R. (1954). Minimizing the number of tankers to meet a fixed schedule. *Naval Research Logistics Quarterly, 1*, 217–222.
Everett, J. L., Hax, A. C., Lewinson, V. A., & Nudds, D. (1972). Optimization of a fleet of large tankers and bulkers: A linear programming approach. *Marine Technology, 20*(1), 430–438.
Fagerholt, K. (1999). Optimal fleet design in a ship routing problem. *International Transactions in Operational Research, 6*(5), 453–464.
Fagerholt, K., & Lindstad, H. (2000). Optimal policies for maintaining a supply service in the Norwegian Sea. *Omega, 28*(3), 269–275.
Gelareh, S., & Meng, Q. (2010). A novel modeling approach for the fleet deployment problem within a short-term planning horizon. *Transportation Research Part E, 46*(1), 76–89.
Laderman, J., Gleiberman, L., & Egan, J. F. (1966). Vessel allocation by linear programming. *Naval Research Logistics Quarterly, 13*(3), 315–320.
Lane, D. E., Heaver, T. D., & Uyeno, D. (1987). Planning and scheduling for efficiency in liner shipping. *Maritime Policy and Management, 12*(3), 109–125.
Mourão, M. C., Pato, M. V., & Paixão, A. C. (2001). Ship assignment with hub and spoke constraints. *Maritime Policy and Management, 29*(2), 135–150.
Papadakis, N. A., & Perakis, A. N. (1989). A nonlinear approach to the multiorigin, multidestination fleet deployment problem. *Naval Research Logistics, 36*, 515–528.
Perakis, A. N. (1985). A second look at fleet deployment. *Maritime Policy and Management, 12*(3), 209–214.
Perakis, A. N., & Jarammillo, D. I. (1991). Fleet deployment optimization for liner shipping Part 1. Background, problem formulation and solution approaches. *Maritime Policy and Management, 18*(3), 183–200.
Perakis, A. N., & Papadakis, N. (1987a). Fleet deployment optimization models Part 1. *Maritime Policy and Management, 14*(2), 127–144.
Perakis, A. N., & Papadakis, N. (1987b). Fleet deployment optimization models Part 2. *Maritime Policy and Management, 14*(2), 145–155.
Powell, B. J., & Perakis, A. N. (1997). Fleet deployment optimization for liner shipping: An integer programming model. *Maritime Policy and Management, 24*(2), 183–192.
Sambracos, E., Paravantis, J. A., Tarantilis, C. D., & Kiranoudis, C. T. (2004). Dispatching of small containers via coastal freight liners: The case of the Aegean Sea. *European Journal of Operational Research, 152*(2), 365–381.
Shintani, K., Imai, A., Nishimura, E., & Papadimitriou, S. (2007). The container shipping network design problem with empty container repositioning. *Transportation Research Part E, 43*(1), 39–59.

CHAPTER NINE

Liner Ship Fleet Planning Problem With a Joint Chance-Constrained Service Level

Contents

INTRODUCTION

The container shipment demand between any two ports of call is an essential input of the LSFD problems. Before the actual container demand is realized, decisions of types and numbers of ships assigned to shipping routes have to be made using the forecasted or estimated container demand. However, some uncontrollable and unpredicted factors, such as the cancelation of a shipping contract or the delay in arrival of containers at the port, do exist in practice. As a result, it is almost impossible to ensure that the estimated container demand matches the realistic demand precisely. Whatever is overestimated or underestimated regarding the demand, it

Liner Ship Fleet Planning
http://dx.doi.org/10.1016/B978-0-12-811502-2.00009-0

will lead to a loss for a liner container shipping company. The potential of uncontrollable and unpredicted factors would result in uncertainty of container shipment demand.

To handle demand uncertainty, Chapter 8 proposed a chance-constrained programming (CCP) approach by which a deterministic LSFD problem was extended to account for the uncertainties. However, the model proposed in this study is based on the assumption that the demand of all port pairs are independent and follow normal distribution without verification, which is not consistent with practice. Therefore this chapter will extend the work in Chapter 8 without such a restriction that the demand of all port pairs are independent and follow a normal distribution. This modeling approach in this chapter is more practical and relevant, as it provides a liner shipping company service information regarding the whole network.

MODEL DEVELOPMENT

As this chapter is an extension of the work in this chapter and we have described the LSFD problem already, here we just simply state it as follows: Determine the number of ships of each type to charter in/out, the type and number of ships to deploy on each shipping route, and the number of voyages to be completed on each shipping route in order to maintain a level of shipping service for shippers while minimizing the total costs.

Before the development of a mathematical programming model for the short-term LSFP problem with a container shipment demand uncertainty, we firstly introduce the decision variables shown as follows:

n_{kr}^{TOTAL}, number of ships (the sum of owned and chartered in ships) of type k ($k \in \mathcal{K}$) assigned on route r ($r \in \mathcal{R}$)

n_{kr}^{IN}, number of chartered in ships of type k ($k \in \mathcal{K}$) assigned on route r ($r \in \mathcal{R}$)

x_{kr}, number of voyages of ships of type k ($k \in \mathcal{K}$) on route r ($r \in \mathcal{R}$).

Chance Constraints for Joint Service Level

Let $\xi^{(p_r^i, p_r^j)}$ be the random variable representing the container shipment demand of a port pair $(p_r^i, p_r^j) \in \mathcal{M}_r$. The container shipment flow on leg l ($l = 1,\ldots,m_r$) of route $r \in \mathcal{R}$, denoted by η_l^r, is given by:

$$\eta_l^r = \sum_{(p_r^i, p_r^j) \in \mathcal{M}_r} \rho_l^{(p_r^i, p_r^j)} \xi^{(p_r^i, p_r^j)}, \quad l = 1, \ldots, m_r; \forall r \in \mathcal{R}. \tag{9.1}$$

Let α_r denote the confidence parameter on route r; therefore, the constraints that the transportation capacity of containerships operated on this route is not less than each container shipment flow on leg l with a least probability of $1 - \alpha_r$ can be formulated as the following chance constraints:

$$\Pr\left(\sum_{k \in \mathcal{K}} x_{kr} V_k \geq \eta_l^r\right) \geq 1 - \alpha_r, \quad l = 1, \ldots, m_r; \forall r \in \mathcal{R}. \tag{9.2}$$

The chance constraints (9.2) are for the individual service level on each shipping route and represents the probability that the shippers' requirements can be guaranteed on each shipping route. Here, we would revise them as a single constraint to represent the service level for the whole shipping network.

Let $\eta^r = \max_{l=1,\cdots,m_r} (\eta_l^r), \forall r \in \mathcal{R}$ to denote the maximal container shipment flow on shipping route r. The liner container shipping company can satisfy the customers' shipping requirement with a probability of $1 - \alpha$ can be formulated as the following probabilistic form, which is termed as a joint chance constraint:

$$\Pr\left(\sum_{k \in \mathcal{K}} x_{kr} V_k \geq \eta^r, \forall r \in \mathcal{R}\right) \geq 1 - \alpha \tag{9.3}$$

Total Costs Function

The total costs of ships incurred in the LSFD problem contain three components: operating costs, voyage costs, and chartering costs; thus we first compute the operating costs. Let c_k^{OPERATE} (USD/day) denote the operating costs of a ship of type $k \in \mathcal{K}$, T (days) denote the length of the short-term planning horizon, andthen the operating costs of all ships in the planning horizon can be computed by $\sum_{r \in \mathcal{R}} \sum_{k \in \mathcal{K}} n_{kr}^{\text{TOTAL}} c_k^{\text{OPERATE}} T$. As for the voyage costs of all ships, it equals $\sum_{r \in \mathcal{R}} \sum_{k \in \mathcal{K}} c_{kr}^{\text{VOYAGE}} x_{kr}$, where c_{kr}^{VOYAGE} (USD/voyage) denote the voyage costs of a ship of type $k \in \mathcal{K}$ on shipping route $r \in \mathcal{R}$. Let c_k^{IN} (USD/day) denote the daily cost of chartering in a ship of type $k \in \mathcal{K}$ for the planning horizon, then the total chartering costs can be computed by $\sum_{k\in\mathcal{K}} n_k^{\text{IN}} c_k^{\text{IN}} T$. For the sake of presentation, we let x be the vector of all decision variables, namely, $\mathbf{x} = \left(n_{kr}^{\text{TOTAL}}, n_k^{\text{IN}}, x_{kr} | k \in \mathcal{K}, r \in \mathcal{R}\right)$. Therefore the function of the total costs of ships incurred in the proposed LSFD problem with respect to the decision vector **x**, denoted by $TC(\mathbf{x})$, equals

$$TC(\mathbf{x}) = \sum_{r\in\mathcal{R}}\sum_{k\in\mathcal{K}} n_{kr}^{\text{TOTAL}} c_k^{\text{OPERATE}} T + \sum_{r\in\mathcal{R}}\sum_{k\in\mathcal{K}} c_{kr}^{\text{VOYAGE}} x_{kr} + \sum_{k\in\mathcal{K}} n_k^{\text{IN}} c_k^{\text{IN}} T, \tag{9.4}$$

Joint Chance-Constrained Programming Model

The proposed short-term LSFP problem with an uncertain container shipment demand can be formulated as the CCP model, named M1:

$$[\text{M1}] \quad z_\alpha^* = \min_{\mathbf{x}} TC(\mathbf{x}), \tag{9.5}$$

subject to:

$$\sum_{r\in\mathcal{R}} n_{kr}^{\text{TOTAL}} \le N_k^{\text{MAX}} + n_k^{\text{IN}}, \forall k \in \mathcal{K} \tag{9.6}$$

$$n_k^{\text{IN}} \le NCI_k^{\text{MAX}}, \forall k \in \mathcal{K} \tag{9.7}$$

$$x_{kr} \le n_{kr}^{\text{TOTAL}} \times \left\lfloor \frac{T}{t_{kr}} \right\rfloor, \forall r \in \mathcal{R}, k \in \mathcal{K} \tag{9.8}$$

$$\sum_{k\in\mathcal{K}} x_{kr} \ge N_r, \forall r \in \mathcal{R} \tag{9.9}$$

$$\Pr\left(\sum_{k\in\mathcal{K}} x_{kr} V_k \ge \eta^r, \forall r \in \mathcal{R}\right) \ge 1 - \alpha \tag{9.10}$$

$$n_{kr}^{\text{TOTAL}}, n_k^{\text{IN}} \quad x_{kr} \in \mathbb{Z}^+ \cup \{0\}, \forall k \in \mathcal{K}, \forall r \in \mathcal{R}, \tag{9.11}$$

where z_α^* denotes the value of the objective function in Eq. (9.5), t_{kr} is the voyage time of a ship of type k on a particular shipping route r (in days), and N_r is the minimal number of voyages required on shipping route r during the planning horizon in order to maintain a given liner shipping service frequency.

Eq. (9.5) is the objective function of the M1 model. The set of constraints (9.6) ensures that the total number of ships used in the fleet, including those owned and those chartered in, does not exceed the number of available ships. The set of constraints (9.7) indicates that the number of chartered-in ships is finite and does not exceed the number of available ships. The right-hand side of constraints (9.8) gives the maximal number of voyages that ships deployed on route r can complete in the planning horizon, where $\lfloor a \rfloor$ denotes the maximum integer not greater than a. Therefore the set of constraints (9.8) is the upper bound for the decision variables x_{kr}. The constraints given by Eq. (9.9) require that ships deployed on shipping route r have to complete at least Nr voyages in order to maintain the given liner shipping frequency. For example, if a weekly shipping service is required on shipping route r during a planning horizon of 6 months, then $N_r = 26$. Constraint (9.10) is a joint chance constraint to define the level of service at the network as $1-\alpha$; namely the ships on all shipping routes can at least satisfy the customers' requirements with a probability of $1-\alpha$.

Constraint (9.10) can be rewritten in another form. Let $G^r(\hat{\mathbf{x}}, \eta^r) := \eta^r - \sum_{k \in \mathcal{K}} x_{kr} V_k$, where $\hat{\mathbf{x}} = (x_{kr} | k \in \mathcal{K}, r \in \mathcal{R})$, and let $G(\hat{\mathbf{x}}, \eta) := \max_{\forall r \in \mathcal{R}} G^r(\hat{\mathbf{x}}, \eta^r)$. We define the probability function $p(\hat{\mathbf{x}}) := \Pr(G(\hat{\mathbf{x}}, \eta) > 0)$, so constraint (9.10) is equivalent to the equation below:

$$p(\hat{\mathbf{x}}) \leq \alpha \tag{9.12}$$

Therefore we have another JCCP model with a joint chance constraint (9.10) replaced by Eq. (9.12), named M2:

$$[\text{M2}] \quad z_\alpha^* = \min_{\mathbf{x}} TC(\mathbf{x}), \tag{9.13}$$

subject to Eqs. (9.6)–(9.9) and (9.12).

SOLUTION ALGORITHM

CCP has been studied extensively in the stochastic programming literature (Prékopa, 2003). However, this problem is still considered challenging because of the two major extreme difficulties to solve it: one is that the feasible region defined by a probabilistic constraint in CCP is generally not convex; another is the chance constraints generally have no closed forms and are typically difficult to evaluate (Miller & Wagner, 1965). To address these difficulties, different approaches have been proposed in the stochastic optimization literature and can be classified into two somewhat different directions: one is to employ convex approximations of chance constraints (Ben-Tal & Nemirovski, 2000; Hong et al., 2011), another is to discretize the probability distribution and use a Monte Carlo simulation to approximate the obtained problem (Dentcheva et al., 2000; Pagnoncelli et al., 2009). The convex approximation approaches usually require that the decision variables are continuous; however, the decision variables involved in the proposed JCCP models (M1 and M2) are restricted to be integers, the convex approximation approaches are thus not applicable for our problem. Therefore the approach in the second direction, specifically, the Sample Average Approximation (SAA) approach (Luedtke & Ahmed, 2008) introduced in Chapter 6, is then used to seek approximation for the proposed JCCP models.

Sample Average Approximation

The theoretical background of SAA approach suggested by Luedtke and Ahmed (2008) is based on the Law of Large Numbers theory, which indicates that the probability of an event occurrence can be approximated by the frequency of the events that occur in number of trials (e.g., S trials). Accordingly, we first let $\xi_1^{(p_r^i,\, p_r^j)} \ldots \xi_S^{(p_r^i,\, p_r^j)}$ be an independent Monte Carlo sample of S realizations of the random variable $\xi^{(p_r^i,\, p_r^j)}$ $\left(\forall\left(p_r^i, p_r^j\right) \in \mathcal{M}_r, \forall r \in \mathcal{R}\right)$, we then obtain the S realization of the random vector η, denoted by $\eta_1, \ldots, \eta_S$. Based on the Law of Large Numbers theory, the probability that $G(\hat{\mathbf{x}}, \eta)$ is larger than 0 will be approximated by the proportion of the count that $G(\hat{\mathbf{x}}, \eta_i) > 0$ appeared in the number of S realizations $(i = 1, \ldots, S)$. Mathematically, let $1_{(0,\, \infty)} : \mathbb{R} \rightarrow \mathbb{R}$ be the indicator function of $(0, \infty)$; that is,

$$1_{(0,\, \infty)}(y) := \begin{cases} 1, & \text{if } y > 0, \\ 0, & \text{if } y \leq 0. \end{cases} \tag{9.14}$$

Then the sample version of the probability function $p(\hat{\mathbf{x}})$ is defined as

$$p^S(\hat{\mathbf{x}}) = S^{-1} \sum_{i=1}^{S} 1_{(0,\, \infty)}(G(\hat{\mathbf{x}}, \eta_i)) \tag{9.15}$$

That is, $p^S(\hat{\mathbf{x}})$ is equal to the proportion of times that $G(\hat{\mathbf{x}}, \eta_i) > 0$. Following Luedtke and Ahmed (2008), the constraint (9.16) is then replaced by

$$p^S(\hat{\mathbf{x}}) \leq \beta, \tag{9.16}$$

where $\beta \in (0, 1)$ is a confidence parameter and can be different from the original one α (the rationale that the confidence parameter in Eq. (9.16) is denoted by β not α is explained in following sections). Finally, the sample version of Model M2 with a joint chance constraint is named M3 and defined as

$$[\text{M3}] \quad \hat{z}_{\beta}^{S} = \min_{\mathbf{x}} TC(\mathbf{x}), \tag{9.17}$$

subject to Eqs. (9.6)–(9.9) and (9.16).

Solving the SAA Problem

According to the above description, the joint chance constraint shown in Eq. (9.10) is replaced by Eq.(9.12), and the original Model M2 is approximated by Model M3; however, it is still hard to solve because of the complexity of constraint (9.16). To solve Model M3, we rewrite it as a mixed-integer linear program (MIP) by introducing one auxiliary variable $\vartheta_i (i = 1, \dots, S)$ for each sample point (Pagnoncelli et al., 2009):

$$[\text{M4}] \quad \hat{z}_{\beta}^{S} = \min_{\mathbf{x}} TC(\mathbf{x}), \tag{9.18}$$

subject to Eqs. (9.6)–(9.9), and

$$\vartheta_i \eta_i^r + \sum_{k \in \mathcal{K}} x_{kr} V_k \geq \eta_i^r, \quad \forall i = 1, \dots, S; \ \forall r \in \mathcal{R} \tag{9.19}$$

$$\sum_{i=1}^{S} \vartheta_i \leq S\beta \tag{9.20}$$

$$\vartheta_i \in \{0, 1\}^S. \tag{9.21}$$

Proposition 9.1. It is noted that Models M3 and M4 are equivalent; see the proof shown in Appendix.

Proposition 9.2. The Models M3 and M4 are equivalent.

Proof. Let $(\mathbf{x}, \vartheta_1, \dots, \vartheta_S)$ be feasible solution for Model M3 shown by Eq. (9.18). For each $i = 1, \dots, S$, from constraints (9.19), we can deduce that if $\forall r \in \mathcal{R}, \sum_{k \in \mathcal{K}} x_{kr} V_k \geq \eta_i^r$, then $\vartheta_i = 0$ or $\vartheta_i = 1$, and we have $1_{(0, \infty)}(G(\hat{\mathbf{x}}, \eta_i)) = 0$; if $\exists r \in \mathcal{R}, \sum_{k \in \mathcal{K}} x_{kr} V_k \geq \eta_i^r$, then $\vartheta_i = 1$

and $1_{(0,\infty)}(G(\hat{\mathbf{x}},\eta_i))=1$; if $\forall r\in\mathcal{R},\sum_{k\in\mathcal{K}}x_{kr}V_k\leq\eta_i^r$, then $\vartheta_i=1$; and $1_{(0,\infty)}(G(\hat{\mathbf{x}},\eta_i))=1$. Therefore $\vartheta_i\geq 1_{(0,\infty)}(G(\hat{\mathbf{x}},\eta_i))$. Accordingly, from constraint (9.20), we have $\beta\geq S^{-1}\sum_{i=1}^{S}\vartheta_i\geq S^{-1}\sum_{i=1}^{S}1_{(0,\infty)}(G(\hat{\mathbf{x}},\eta_i))=p^S(\hat{\mathbf{x}})$. Thus, x is feasible to Eq. (9.17) and has the same objective value as in Eq. (9.18). Conversely, let x be a feasible solution for Eq. (9.17), and define $\vartheta_i=1_{(0,\infty)}(G(\hat{\mathbf{x}},\eta_i))$. For each $i=1,\ldots,S$, if $G(\hat{\mathbf{x}},\eta_i)\leq 0$, then $\vartheta_i=0$ and $\sum_{k\in\mathcal{K}}x_{kr}V_k>\eta_i^r$, $\forall r\in\mathcal{R}$; therefore constraints (9.19) hold; if $G(\hat{\mathbf{x}},\eta_i)>0$, then $\vartheta_i=1$ and constraint (9.19) holds as well. As for constraint (9.20), we have $\sum_{i=1}^{S}\vartheta_i=\sum_{i=1}^{S}1_{(0,\infty)}(G(\hat{\mathbf{x}},\eta_i))\leq S\beta$. Therefore, we have that $(\mathbf{x},\vartheta_1,\ldots,\vartheta_S)$ is feasible for Model M4 expressed by Eqs. (9.18)–(9.21) with the same objective values. The proposition is proved.□

Since Model M4 is an MIP, the optimization solver CPLEX can be employed to solve it. Therefore we can state that solving the original model M2 has been converted into solving the model M4. Let $\mathbf{X}_\alpha$ and $\mathbf{X}_\beta^S$ denote the set of optimal solutions to the true problem (i.e., Model M2) and the SAA problem (i.e., Model M4), respectively. It has been proved that $\hat{z}_\beta^S$ and $\mathbf{X}_\beta^S$ converge w.p.1 to their counterparts of the true problem (i.e., z_α^* and $\mathbf{X}_\alpha$) exponentially fast as S increases under mild regularity conditions (Pagnoncelli et al., 2009), which justifies the solution algorithm described above and indicates that the solution quality of Model M4 is guaranteed.

Lower Bound

Increasing the feasible set of an optimization problem aiming at minimizing the value of an objective function may result in a decrease in the optimal objective function value of the problem. Therefore if we increase the value of α in Model M2, then z_α^* may decrease. In other words, we can obtain a lower bound of M2 by increasing the value of α. However, solving M2 is extremely difficult, which indicates that it is hard for us to obtain the lower bound by solving M2 with an enlarged α. As Model M4 is an approximation of Model M2, we can expect that the objective function value of Model M4 in which $\beta>\alpha$, denoted by $\hat{z}_{\beta L}^S$, is a lower bound of z_α^* with some significance level. This expectation has been mathematically proved in Theorem 3 of Luedtke and Ahmed (2008); accordingly, the sample size S ensures that $\hat{z}^S_{\beta^L} \leq z^*_\alpha$ with a probability of at least $1-\delta$, where $\delta \in (0, 1)$, can be estimated by

$$S \geq \frac{1}{2(\beta-\alpha)^2} \ln\left(\frac{1}{\delta}\right), \tag{9.22}$$

Verification of Solution Feasibility

The previous section shows that solving Model M4, in which $\beta^L > \alpha$ yields a lower bound of Model M2 with some probability. Contrarily, solving Model M4 with $\beta^U < \alpha$ might produce feasible solutions to Model M2. In other words, it yields an upper bound with some probability, denoted by $\hat{z}^S_{\beta^U}$. For a given candidate point $\mathbf{x}^* \in \mathbf{X}^S_{\beta^U}$, namely an optimal solution to Model M4 in which $\beta^U < \alpha$, we have to validate its quality as a solution to Model M2. For that, we need to estimate the probability $p(\hat{\mathbf{x}}^*)$, and so the Monte Carlo sampling technique is thus employed again. The procedure is similar to that described above namely, independently generate S' realizations of for random variables $\xi_1^{(p_r^i, p_r^j)} \ldots \xi_{S'}^{(p_r^i, p_r^j)}$ and then estimate $p(\hat{\mathbf{x}}^*)$ by $p^{S'}(\hat{\mathbf{x}}^*)$ because the estimator $p^{S'}(\hat{\mathbf{x}}^*)$ is unbiased. It is noted that we can use a very large sample S', as there is no need to solve any optimization problem here. If $p^{S'}(\hat{\mathbf{x}}^*) \leq \alpha$, then $\mathbf{x}^*$ is a feasible solution. Otherwise, we choose another smaller β^U, obtain a new solution $\mathbf{x}^*$, and check its feasibility. This procedure is repeated until a feasible solution is obtained. It should be mentioned that our computational experiments actually demonstrate that a feasible solution is generally obtained in the first iteration.

NUMERICAL EXAMPLE

In this section, we first conduct a sensitivity analysis of SAA parameters through a preliminary experiment with small scales. This is done in order to choose suitable values of SAA parameters while taking into account the trade-off between the quality of the solution obtained for the experiment and the computational effort needed to solve it. With these chosen parameters, we then illustrate the applicability of the proposed methodology and conduct risk management on a real-world shipping network. The solution algorithm is implemented in a programming language Lua (v5.1) coded in

Microcity (http://microcity.sourceforge.net), and the SAA problems are solved by CPLEX (v12.1). All computations are carried out on a desktop personal computer with Intel Core 2 CPU 3.10 GHz and 4.0 GB of RAM under Microsoft Windows XP.

Sensitivity Analysis of SAA Parameters

From the above description of the SAA approach, it is found that for a JCCP problem with a given confidence parameter α, the parameters, β, δ, S, and S' need to be determined in the SAA approach. Therefore, the sensitivity analysis of SAA parameters focuses on β, δ, and S, and it is implemented like this: We first test a number of sets of these SAA parameters; the results are shown in Table 9.1. We then evaluate the performance of the approach with these tested SAA parameters in order to choose the best one.

We set three different values of α shown in the first column of Table 9.1. For each value of α, five sets of parameters, β, δ, S, and S', are tested. The values of S in the fifth column satisfy Eq. (9.22). The relative gap between the lower bound and upper bound is computed by $\hat{z}^S_{\beta^U} - \hat{z}^S_{\beta^L}/\hat{z}^S_{\beta^L} \times 100\%$, shown in Column 7. The computational time is listed in the last column of Table 9.1.

Table 9.1 Sensitivity Analysis of SAA Parameters

α	β^L	β^U	δ	S	S'	Relative Gap (%)	$p^{S'}(\hat{x}^*)$	CPU Time (s)
0.05	0.08	0.030	0.10	1800	10,000	0.85	0.0155	1695.52
	0.10	0.025	0.09	500	8000	0.73	0.0260	58.91
	0.12	0.020	0.08	300	6000	0.85	0.0162	30.99
	0.15	0.015	0.07	200	4000	1.24	0.0171	14.53
	0.20	0.010	0.06	100	2000	1.36	0.0092	2.66
0.10	0.12	0.075	0.095	3000	10,000	0.37	0.0795	587.03
	0.15	0.070	0.09	500	8000	0.76	0.0521	15.57
	0.18	0.060	0.08	200	6000	0.76	0.0468	8.23
	0.20	0.055	0.07	150	4000	0.76	0.0526	3.89
	0.25	0.050	0.06	100	2000	0.88	0.0415	3.63
0.15	0.18	0.125	0.06	1600	10,000	0.39	0.0755	65.94
	0.20	0.100	0.05	600	8000	0.39	0.1293	4.19
	0.25	0.075	0.04	200	6000	0.76	0.0494	0.60
	0.28	0.050	0.03	120	4000	0.94	0.0260	0.57
	0.30	0.025	0.02	100	2000	1.18	0.0271	0.47

As can be seen from Table 9.1, for each α, the relative gap generally increases with the interval between β^L and β^U. The rationale behind this trend is that when β^L increases, the feasible set increases as well, which results in a possible decrease in the lower bound $\hat{z}^S_{\beta^L}$. Similarly the upper bound $\hat{z}^S_{\beta^U}$ may increase when β^U decreases. Therefore it makes the relative gap enlarge for an increasing interval between β^L and β^U. However, an exception in Table 9.1 is that the first relative gap in Column 8 for $\alpha=0.05$ is 0.85%, which is larger than the second one, 0.73%. It is possible for this exception because the JCCP models (or SAA models) involve uncertain parameters, and their values are generated randomly. The randomness of parameters may make such an exception occur. Additionally, all of the values of $p^{S'}(\hat{\mathbf{x}}^*)$ in Column 8 are less than the corresponding value of α, which indicates that the values set for β^U and S' are effective in yielding a feasible solution.

Computational Results

Here we still use the numerical example taken in Chapter 8. The liner ship route network consisting of eight routes operated by a liner container shipping company, OOCL in Hong Kong (see Fig. 8.2), and the relevant ship data is shown in Table 8.3. We assume that the uncertain parameters of the container demand in the LSFD test experiment follow log-normal distributions, such as $\xi^{(p_r^i, p_r^j)} \sim \ln N\left(\mu^{(p_r^i, p_r^j)}, \sigma^{(p_r^i, p_r^j)^2}\right)$, to generate the demands. This is because log-normal distributions are well suited for modeling random variables such as uncertain demands (Kamath & Pakkala, 2002). For the sake of presentation, the ratio $\sigma^{(p_r^i, p_r^j)}/\mu^{(p_r^i, p_r^j)}$ is assumed to be the same for all port pairs, denoted by λ. Assuming that $\alpha=0.10$, we set $\beta^L=0.12$, $\beta^U=0.075$, $\delta=0.095$, $S=3000$, and $S'=10{,}000$ for the following analysis.

The variance of an uncertain container demand represents the risk of shipping market. In the case where $\sigma^{(p_r^i, p_r^j)}=0$, the shipping market can be seen as having a lower risk, and the container demand can be predicted precisely. In the case where $\sigma^{(p_r^i, p_r^j)}>0$, it indicates that there is risk in the shipping market; when $\sigma^{(p_r^i, p_r^j)}$ increases, it means that the risk increases as well. In order to study the effect of the variance on the costs that the liner shipping company must take on in order to maintain a given level of service (i.e., the objective function value of the JCCP model), we vary the ratio λ

from 0 to 0.5 with increments of 0.05 and show the trend in the cost as λ changes in Fig. 9.1. As can be seen, the trend in cost generally increases with λ increases. It shows that the variability of the uncertain parameters has a significant effect on the solutions. The assignment of types and numbers of ships to each shipping route and the voyages on each of those routes are the output of CPLEX optimization solver, which can be adapted in practice by the liner container shipping company. For example, Table 9.2 lists the outputs to the LSFD problem of the case $\lambda=0.1$. Due to space limitation the outputs of the cases when λ varies from 0 to 0.5 with increments of 0.05 are not shown here.

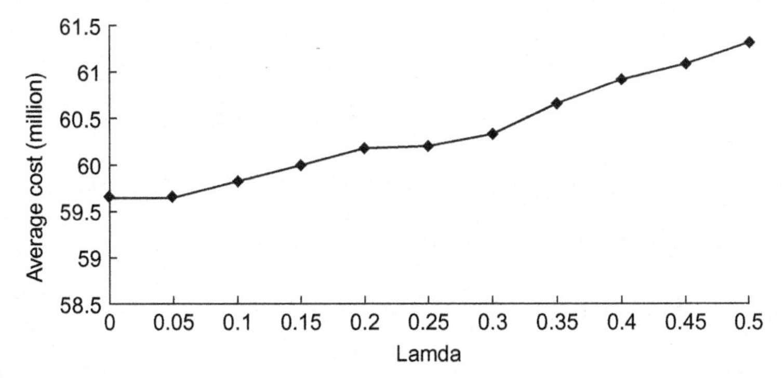

Fig. 9.1 Average cost for different levels of variance.

Table 9.2 Results of Ship Assignment for the Case $\lambda=0.1$

	Route Type							
Ship Type	**CCX**	**CPX**	**GIS**	**IDX**	**NCE**	**NZX**	**SCE**	**UKX**
Ship allocations								
1	3						4	
2		2	1			1	3	
3			1	5	8			
4						2		
5	1	1		1	1		2	1
Number of voyages								
1	21						12	
2		18	13			8	9	
3			13	20	24			
4						18		
5	5	8		6	2		6	26

SUMMARY

In this chapter, we considered a realistic LSFP problem with container demand uncertainty that was encountered by a liner shipping company. A concept of level of service is introduced in this problem to deal with managing the risk of and uncertain container demand; a JCCP model is proposed for it. The contribution of this study is fourfold: first, it contributes to the literature by proposing a realistic LSFD problem with uncertain container demand; second, the proposed LSFD problem is an extension of the study of Meng and Wang (2010) and formulated as a JCCP model; third, an appropriate solution algorithm is proposed to solve the JCCP model; and fourth, which is perhaps the most significant, the methodology of model development and solution algorithm design could be also applied to other similar problems under uncertain environments, such as an air transportation planning problem with uncertain demand, etc.

The challenge to solve the JCCP model proposed in this chapter is that the joint chance constraints generally have no closed forms and are thus hard to evaluate. To effectively solve the proposed JCCP model, we first proposed an SAA model to approximate the model; further, we transformed the SAA model into an equivalent MIP model and solved it by using a CPLEX solver. A sensitivity analysis of SAA parameters through a preliminary experiment was first conducted, then the proposed model and solution algorithm were tested using a real world liner shipping network. The gaps between the lower and upper bounds are small, which indicates that the solution scheme is effective. It was also found that the variability of the uncertain parameters has a significant effect on the solutions. The study of this chapter has been published in a journal already; readers who are interested in can refer to Wang et al. (2013).

REFERENCES

Ben-Tal, A., & Nemirovski, A. (2000). Robust solutions of linear programming problems contaminated with uncertain data. *Mathematical Programming*, *88*(3), 411–424.

Dentcheva, D., Prékopa, A., & Ruszczyński, A. (2000). Concavity and efficient points of discrete distribution in probabilistic programming. *Mathematical Programming*, *89*(1), 55–77.

Hong, L., Yang, Y., & Zhang, L. (2011). Sequential convex approximation to joint chance constrained programs: A Monte Carlo approach. *Operations Research*, *59*(3), 617–630.

Kamath, K., & Pakkala, T. (2002). A Bayesian approach to a dynamic inventory model under an unknown demand distribution. *Computers and Operations Research*, *29*(4), 403–422.

Luedtke, L., & Ahmed, S. (2008). A sample approximation approach for optimization with probalistic constraints. *SIAM Journal on Optimization*, *19*(2), 674–699.

Meng, Q., & Wang, T. (2010). A chance constrained programming model for short-term liner ship fleet planning problems. *Maritime Policy and Management*, *37*(4), 329–346.

Miller, L., & Wagner, H. (1965). Chance-constrained programming with joint constraints. *Operations Research*, *13*(6), 930–945.

Pagnoncelli, B., Ahmed, S., & Shapiro, A. (2009). Sample average approximation method for chance constrained programming: Theory and applications. *Journal of Optimization Theory and Applications*, *142*(2), 399–416.

Prékopa, A. (2003). Probability programming. In A. Ruszczyński & A. Shapiro (Eds.), *Handbook in OR & MS: Vol. 10. Stochastic programming* (pp. 267–352). Amsterdam: Elsevier

Wang, T., Meng, Q., Wang, S., & Tan, Z. (2013). Risk management in liner ship fleet deployment: A joint chance constrained programming model. *Transportation Research Part E*, *60*, 1–12.

CHAPTER TEN

Liner Ship Fleet Planning with Expected Profit Maximization

Contents

INTRODUCTION

Nowadays, transshipping containers at a hub port are typical liner shipping operations because they deploy large ships calling at hub ports to benefit the economies of scale in ship size (Cullinane & Khanna, 1999). As reported by Vernimmen, Dullaert, and Engelen (2007), about one-third of the laden container throughput in the world is made up of transshipped containers. Mourão, Pato, and Paixão (2001) made the first attempt to solve the liner ship fleet deployment problem with a container transshipment and deterministic container shipment demand. They investigated a hypothetical hub-and-spoke (H&S) network with one pair of ports and two ship routes: one feeder route and one main route. All containers had to be transshipped at the hub port in the feeder route. However, this model is too simple to reflect the realistic ship fleet deployment, therefore this chapter studies the short-term LSFP problem with container transshipment and uncertain container shipment demand.

Liner Ship Fleet Planning
http://dx.doi.org/10.1016/B978-0-12-811502-2.00010-7

Container transshipment operations mean that there can be multiple container routes between an origin and a destination, and some of these container routes involve more than one ship route. The short-term LSFP problem should therefore choose the best container routes and assign the right number of containers to each of these container routes. In fact, though, the short-term LSFP problem takes into account container transshipment, and demand uncertainty is a new research issue with practical importance. Most of the existing relevant literature (Christiansen, Fagerholt, & Ronen, 2004; Perakis, 2002; Ronen, 1983, 1993) assumes deterministic container shipment demand. This chapter thus focuses on model formulation and algorithm development for this new research issue.

In this chapter, we investigate the short-term LSFP problem with container transshipment and uncertain container shipment demand. To characterize the uncertainty, we first assume that the number of containers transported from an origin port to a destination port is a random variable. With these random container shipment demands, the proposed LSFP problem can be formulated as a two-stage stochastic integer programming model with the objective of maximizing the expected value of the total profit. To solve this a solution algorithm integrating the sample average approximation method and a dual decomposition and Lagrangian relaxation approach will be developed.

MODEL DEVELOPMENT

Before the development of a two-stage stochastic programming model, which aims to maximize the expected profit for the short-term LSFP problem with container transshipment and uncertain container shipment demand, we firstly introduce the concept of container route to deal with the container transshipment issue.

Container Routes With Container Transshipment Operations

Let $\mathcal{W} = \{(o, d) | o \in \mathcal{P}, d \in \mathcal{P}\}$ be the set of origin-to-destination (O-D) port pairs with container shipment demand, and ξ^{od} e the number of containers in terms of TEUs (i.e., 20-foot equivalent units) to be transported between an O-D port pair $(o, d) \in \mathcal{W}$ in the short-term planning horizon (six months). As aforementioned in Chapter 8, the liner container shipping company provides regular shipping service on a predetermined liner ship route network; in other words, the route of ships is fixed (the routes are listed in Table 8.1). However, the route of containers may be different from the route of ships because there are usually many candidate routes for transporting containers from their origin to their destination due to transshipment. Given the set of ship routes $\mathcal{R}$ the liner container shipping company can predetermine a set of candidate container routes to deliver containers between an O-D port pair $(o, d) \in \mathcal{W}$, denoted by $\mathcal{H}^{od}$. The container route $h^{od} \in \mathcal{H}^{od}$ is either a part of one particular ship route or a combination of several ship routes to deliver containers from original port $o \in \mathcal{P}$ to the destination port $d \in \mathcal{P}$. Container transshipment operations are involved in any container route made up of several ship routes. For example, there are two possible container routes from Jakarta (JK) to Shanghai (SH) in Fig. 10.1:

$$h_1^{\mathrm{JK}\cdot\mathrm{SH}} = p_1^1(\mathrm{JK}) \xrightarrow{\text{Ship route 1}} p_1^2(\mathrm{SG}) \mapsto p_3^2(\mathrm{SG}) \xrightarrow{\text{Ship route 3}} p_3^3(\mathrm{SH}) \tag{10.1}$$

$$h_2^{\mathrm{JK}\cdot\mathrm{SH}} = p_2^1(\mathrm{JK}) \xrightarrow{\text{Ship route 2}} p_2^2(\mathrm{SH}). \tag{10.2}$$

The first container route $h_1^{\mathrm{JK}\cdot\mathrm{SH}}$, made up of two ship routes, involves container transshipment operations: Containers are loaded at the first port call of ship route 1 (Jakarta) and delivered to the second port of call of ship route 3 (Singapore). At the Singapore port, these containers are discharged and reloaded (transshipped) to a ship deployed on ship route 3 and transported to the destination port, Shanghai. However, the second container route $h_2^{\mathrm{JK}\cdot\mathrm{SH}}$ provides direct delivery service via ship route 2 without container transshipment.

A container route contains all of the information on how the containers will be transported, such as origin, destination, ports of call along the route, and any transshipment ports. The introduction of the concept of container

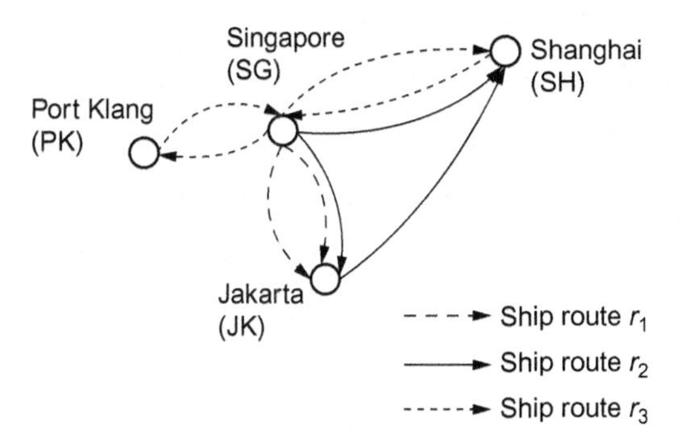

Fig. 10.1 Three-liner ship route.

route facilitates the model formulation as the complex container delivery process is simplified and represented by a finite number of container routes. Some container routes for the liner shipping network in Fig. 10.1 are provided in Table 10.1. An O-D port pair may have several container routes, and the volume of containers to be transported between this O-D port pair could be split among these container routes. Let $\mathcal{H}$ be the set of these entire predetermined container routes for all of the O-D port pairs; that is,

$$\mathcal{H}=\bigcup_{(o,d)\in\mathcal{W}}\mathcal{H}^{od} \tag{10.3}$$

Two-Stage Stochastic Integer Programming Model

Before the development of an optimization model, which aims to maximize the expected value of profit for the short-term LSFP problem with container transshipment and uncertain demand, the following decision variables are introduced as follows:

n_{kr}^{TOTAL}, number of ships (the sum of owned and chartered in ships) of type k deployed on ship route r

n_k^{IN}, number of chartered in ships of type k

n_k^{OUT}, number of chartered out ships of type k

x_{kr}, number of voyages made by ships of type k deployed on ship route r

$z^{h^{od}}$, number of TEUs between an O-D port pair (o, d) transported on container route h^{od}

In the short-term LSFP problem the liner container shipping company not only uses its own ships to deliver containers, but also charters ships from other liner shipping companies. Generally, there are three types of chartering: bareboat charter, voyage charter, and time charter. Bareboat charter is the simplest method, in which the charterer manages the ship and pays all costs except the capital repayment, tax, and depreciation.

Table 10.1 Container Route Plans for Different O-Ds

O-D	**Container Route Plans**
JK-SH	$h_1^{\text{JK}\cdot\text{SH}}$: $p_1^1(\text{JK})\rightarrow p_1^2(\text{SG})/p_3^2(\text{SG})\rightarrow p_3^3(\text{SH})$
	$h_2^{\text{JK}\cdot\text{SH}}$: $p_2^1(\text{JK})\rightarrow p_2^2(\text{SH})$
SH-PK	$h_1^{\text{SH}\cdot\text{PK}}$: $p_3^3(\text{SH})\rightarrow p_3^4(\text{SG})\rightarrow p_3^1(\text{PK})$
SH-SG	$h_1^{\text{SH}\cdot\text{SG}}$: $p_2^2(\text{SH})\rightarrow p_2^3(\text{SG})$
	$h_2^{\text{SH}\cdot\text{SG}}$: $p_3^3(\text{SH})\rightarrow p_3^4(\text{SG})$

In other words the ship owner does not bear any cost except that involved in collecting the rent from the charterer. For simplicity, the bareboat charter is assumed to be the only type included in this study. The rate for chartering a ship of type k from the ship owner over the planning horizon is denoted by c_k^{IN} (\$/ship). Besides paying the chartering rate to the ship owner, the ship charterer incurs other charges in operating the chartered ship, such as routine maintenance costs and insurance. Therefore we have

$$c_k^{\text{OUT}} < c_k^{\text{IN}}, \tag{10.4}$$

where c_k^{OUT} denotes the rate received for chartering out a ship of type k (\$/ship). The revenue obtained by the liner container shipping company has two sources: one from chartering out ships and the other from transporting containers for shippers.

The revenue earned by the liner container shipping company comes from two resources: one is the rent of chartering out ships to other liner operators, and the other is freight rate of shipping containers for shippers. The revenue gained from chartering ships out is given by:

$$\sum_{k \in \mathcal{K}} c_k^{\text{OUT}} n_k^{\text{OUT}}. \tag{10.5}$$

As for the revenue of shipping containers, it is unknown due to the uncertainty of container shipment demand. Let ξ be a random vector defined over a probability space (Ω, F, P), where Ω is the set of elementary outcomes ω, F is the event space, and P is the probability measure. The container shipment demand of an O-D port pair $(o, d) \in \mathcal{W}$ denoted by ξ^{od} is a random variable. Given ω^{od}, which is a realization of the random parameter ξ^{od}, then $z^{h^{od}}$ is obviously a function with respect to ω^{od}. Let f^{od} denote the freight rate of delivering a container with an O-D port pair $(o, d) \in \mathcal{W}$ (\$/TEU). The revenue of shipping containers for all O-D port pairs along all liner ship routes is given by:

$$\sum_{(o,d) \in \mathcal{W}} \sum_{h^{od} \in \mathcal{H}^{od}} f^{od} z^{h^{od}}\left(\omega^{od}\right), \tag{10.6}$$

The total costs incurred by the liner container shipping company consist of three components: container handling cost, ship operating cost, and ship chartering in cost. The container handling cost incurred on a container route, includes the loading cost at the origin port, container discharging cost at the destination port, and transshipment costs at any transshipment ports.

Different container routes between an O-D port pair may result in different container handling costs. For example, the first container route shown in Eq. (10.1) and the second container route shown in Eq. (10.2) both involve the container loading cost at JK and container discharging cost at SH, but the first container route is associated with an additional transshipment cost at SG. Let $c^{h^{od}}$ (\$/TEU) denote the container handling cost per TEU incurred on the container route $h^{od} \in \mathcal{H}^{od}$, and the total container handling cost can be calculate as:

$$\sum_{(o,\, d) \in \mathcal{W}} \sum_{h^{od} \in \mathcal{H}^{od}} c^{h^{od}} z^{h^{od}} \left(\omega^{od}\right). \tag{10.7}$$

Let c_{kr} denote the operating costs for ships of type k on ship route r per voyage, including fuel consumption costs, administration costs, fixed daily operating costs, port charges, and canal fees (if any). The total ship operating cost plus the rent paid for chartering in ships is given by

$$\sum_{r \in \mathcal{R}} \sum_{k \in \mathcal{K}} c_{kr} x_{kr} + \sum_{k \in \mathcal{K}} c_k^{\mathrm{IN}} n_k^{\mathrm{IN}}. \tag{10.8}$$

It should be noted that the decisions regarding n_{kr}^{TOTAL}, n_k^{IN}, n_k^{OUT} and x_{kr} are made prior to the realization of the random container shipment demand ξ, while the decision regarding $z^{h^{od}}$ is determined only after the realization of ξ. We can thus break down the decisions into two stages. In a two-stage stochastic optimization model, the set of decisions are divided into two groups: The first-stage decision variables are those that have to be decided before the actual realization of the uncertain parameters and often referred to as here-and-now decisions. Subsequently, when the random events have presented themselves, further design or operational policy improvements can be made by selecting the values of the second-stage decision variables, which are often referred to as wait-and-see decisions. Therefore in our LSFP problem, the set of decisions is broken down into two groups: n_{kr}^{TOTAL}, n_k^{IN}, n_k^{OUT} and x_{kr} are first-stage decision variables because they are determined before knowing the actual container shipment demand of each O-D port pairs; once they are determined, the number of containers picked up and delivered by ships are then further determined; namely, $z^{h^{od}}$ are second-stage decision variables. As the objective of the two-stage stochastic integer programming (2SSIP) model is to choose the first-stage variables in a way that the sum of the profit of the first stage and the expected profit of the second stage is maximized, the optimization model of the short-term LSFP is given by:

$$\max Z(\widetilde{\mathbf{v}}) = \sum_{k\in\mathcal{K}} c_k^{\text{OUT}} n_k^{\text{OUT}} - \sum_{r\in\mathcal{R}}\sum_{k\in\mathcal{K}} c_{kr}x_{kr} - \sum_{k\in\mathcal{K}} c_k^{\text{IN}} n_k^{\text{IN}} + \mathbb{E}\left[\hat{Q}_{\xi}(\widetilde{\mathbf{v}}, \boldsymbol{\xi})\right], \tag{10.9}$$

subject to

$$\sum_{r\in\mathcal{R}} n_{kr}^{\text{TOTAL}} \le N_k^{\text{MAX}} + NCI_k^{\text{MAX}}, \quad \forall k\in\mathcal{K} \tag{10.10}$$

$$n_k^{\text{IN}} \le NCI_k^{\text{MAX}}, \quad \forall k\in\mathcal{K} \tag{10.11}$$

$$n_k^{\text{OUT}} = N_k^{\text{MAX}} + n_k^{\text{IN}} - \sum_{r\in\mathcal{R}} n_{kr}^{\text{TOTAL}}, \quad \forall k\in\mathcal{K} \tag{10.12}$$

$$x_{kr} \le n_{kr}^{\text{TOTAL}} \times \left\lfloor \frac{T}{t_{kr}} \right\rfloor, \quad \forall r\in\mathcal{R}, k\in\mathcal{K} \tag{10.13}$$

$$\sum_{k\in\mathcal{K}} x_{kr} \ge N_r, \quad \forall r\in\mathcal{R} \tag{10.14}$$

$$n_{kr}^{\text{TOTAL}}, x_{kr} \in \mathbb{Z}^+ \cup \{0\}, \quad \forall k\in\mathcal{K}, \forall r\in\mathcal{R} \tag{10.15}$$

$$n_k^{\text{IN}} \ge 0, \quad \forall k\in\mathcal{K} \tag{10.16}$$

$$n_k^{\text{OUT}} \ge 0, \quad \forall k\in\mathcal{K}, \tag{10.17}$$

where the vector $\widetilde{\mathbf{v}} = \left(\cdots n_{kr}^{\text{TOTAL}} \cdots x_{kr} \cdots n_k^{\text{IN}} \cdots n_k^{\text{OUT}} \cdots\right)$ contains all of the first-stage decision variables, T is the length of planning horizon, t_{kr} is the voyage time of a ship of type k on a particular ship route r (in days), N_r is the minimal number of voyages required on the ship route r during the planning horizon in order to maintain a given liner shipping service frequency. $\mathbb{E}\left[\hat{Q}_{\xi}(\widetilde{\mathbf{v}}, \boldsymbol{\xi})\right]$ is the expected recourse function, in which $\hat{Q}_{\xi}(\widetilde{\mathbf{v}}, \boldsymbol{\xi})$ is the optimal objective function value for the following second-stage optimization problem with a given vector $\widetilde{\mathbf{s}}$ and a random shipment demand ξ. For a particular container shipment demand realization $\xi(\omega)$, we let $\hat{Q}_{\xi}(\widetilde{\mathbf{v}}, \xi(\omega))$ be the value of a maximization problem defined as follows:

$$\hat{Q}_{\xi}(\widetilde{v}, \xi(\omega)) = \max \sum_{(o,d)\in\mathcal{W}} \sum_{h^{od}\in\mathcal{H}^{od}} \left(f^{od} - c^{h^{od}}\right) z^{h^{od}}\left(\xi^{od}(\omega)\right), \tag{10.18}$$

subject to

$$\sum_{k\in\mathcal{K}} x_{kr}V_k \geq \sum_{(o,\,d)\in\mathcal{W}} \sum_{h^{od}\in\mathcal{H}^{od}} \rho_{ir}^{h^{od}} z^{h^{od}}\left(\xi^{od}(\omega)\right),\ \ \forall r\in\mathcal{R}, i=1,\cdots,m_r \quad (10.19)$$

$$\sum_{h^{(o,\,d)}\in\mathcal{H}^{(o,\,d)}} z^{h^{od}}\left(\xi^{od}(\omega)\right) \leq \xi^{od}(\omega),\ \ \forall (o,d)\in\mathcal{W} \quad (10.20)$$

$$z^{h^{od}}\left(\xi^{od}(\omega)\right) \geq 0,\ \ \forall (o,d)\in\mathcal{W},\ \ \forall h^{od}\in\mathcal{H}^{od}, \quad (10.21)$$

where $z^{h^{od}}\left(\xi^{od}(\omega)\right)$ is the number of containers between O-D port pair $(o, d)\in\mathcal{W}$ transported by ships on container route $h^{od}\in\mathcal{H}^{od}$ for realization $\xi^{od}(\omega)$, V_k is the size of a particular ship k (in TEUs), and $\rho_{ir}^{h^{od}}$ is a binary parameter, which equals 1 if a container route $h^{od}\in\mathcal{H}^{od}$ contains leg i of ship route r; otherwise, it is equal to 0.

Eq. (10.9) is the objective function of the two-stage stochastic integer programming model, which is equivalent to maximizing the expected total profit. The constraints given by Eq. (10.10) ensure that the total number of ships used in the fleet, including those owned and those chartered in, does not exceed the number of available ships. The constraints given by Eq. (10.11) indicate that the number of chartered-in ships is finite and does not exceed the number of available ships. The number of chartered-out ships of each type is given by Eqs.(10.12). The constraints given by Eq. (10.13) fix the maximal number of voyages that ships of type k can complete on ship route r, where $\lfloor a\rfloor$ denotes the maximum integer not greater than a. The constraints given by Eq. (10.14) guarantee the number of voyages that are required on ship route r in order to maintain the given liner shipping frequency. For example, if a weekly shipping service is required on ship route r during a planning horizon of six months, then $N_r = 26$. Constraints (10.15) require that n_{kr}^{TOTAL} and x_{kr} are nonnegative integers.

The left-hand sides of constraints (10.19) give the total transportation capacity of ships deployed on the liner ship route $r\in\mathcal{R}$. The right-hand sides give the total number of containers carried by ships sailing on leg i of route $r\in\mathcal{R}$, including those containers loaded at previous ports that have remained on the ships, as well as those loaded or transshipped at port p_r^i. Constraints (10.19) ensure that the container flow of each ship on each leg of each route does not exceed the ship capacity of the ships deployed on that route. Constraints (10.20) imply that the total number of containers assigned to all ship routes between an O-D port pair does not exceed the corresponding container shipment demand realization.

The following proposition indicates that n_k^{OUT} can be relaxed to be a real number, as given by Eq. (10.17).

Proposition 10.1. Assuming that $n_k^{\text{IN}*}$ and $n_k^{\text{OUT}*}$ are respective optimal solutions for n_k^{IN} and n_k^{OUT} in the two-stage stochastic programming model, both $n_k^{\text{IN}*}$ and $n_k^{\text{OUT}*}$ take integer values.

Proof. Suppose that $n_k^{\text{IN}*}$ is not an integer for a particular ship type k. We have

$$\lfloor n_k^{\text{IN}*} \rfloor < n_k^{\text{IN}*}. \tag{10.22}$$

According to Eq. (10.12) and the integrality property of n_{kr}^{TOTAL} and N_k^{MAX}, it follows that

$$\lfloor n_k^{\text{OUT}*} \rfloor < n_k^{\text{OUT}*} \tag{10.23}$$

According to the assumption shown in Eq.(10.4), it is easily seen that the objective function value of model (10.9) will increase by replacing $n_k^{\text{IN}*}$ and $n_k^{\text{OUT}*}$ with $\lfloor n_k^{\text{IN}*} \rfloor$ and $\lfloor n_k^{\text{OUT}*} \rfloor$, respectively, while satisfying all the constraints. This contradicts the optimality of $n_k^{\text{IN}*}$ and $n_k^{\text{OUT}*}$. The integrality of $n_k^{\text{OUT}*}$ can be demonstrated using the same method.□

Remark: Relaxing the integrality constraint on n_k^{IN} and n_k^{OUT} reduces the number of integer variables in the models. As a consequence, the models would require a shorter computational time to be solved.

Substituting n_k^{OUT} in the objective function expressed by Eq. (10.9) with the right-hand sides of Eqs. (10.12) yields the following two-stage stochastic integer programming (2SSP) model with fewer first-stage decision variables denoted by the vector $\mathbf{v} = \left(\cdots n_{kr}^{\text{TOTAL}} \cdots x_{kr} \cdots n_k^{\text{IN}} \cdots\right)$:

[2SSIP]

$$\min Z(\mathbf{v}) = \mathbf{c}^T \mathbf{v} + \mathbb{E}\left[Q_{\xi}(\mathbf{v}, \xi)\right] - \sum_{k \in \mathcal{K}} c_k^{\text{OUT}} N_k^{\text{MAX}}, \tag{10.24}$$

subject to the constraints (10.10), (10.11), (10.13)–(10.16)

$$N_k^{\text{MAX}} + n_k^{\text{IN}} - \sum_{r \in \mathcal{R}} n_{kr}^{\text{TOTAL}} \geq 0, \quad \forall k \in \mathcal{K}, \tag{10.25}$$

where vector $\mathbf{c} = \left(\cdots c_k^{\text{OUT}} \cdots c_{kr} \cdots c_k^{\text{IN}} \cdots\right)$ contains the cost coefficients of the first-stage problem. Let $Q_{\xi}(\mathbf{v}, \xi(\omega))$ be the optimal objective function value for the second-stage optimization problem for a given vector **s** and a given realization $\xi(\omega)$:

$$Q_{\boldsymbol{\xi}}(\mathbf{v}, \xi(\omega)) = \min \sum_{(o,\, d)\in\mathcal{W}} \sum_{h^{od}\in\mathcal{H}^{od}} \left(c^{h^{od}} - f^{od}\right) z^{h^{od}}\left(\xi^{od}(\omega)\right), \qquad (10.26)$$

subject to the constraints (10.19)–(10.21).

SOLUTION ALGORITHM

The 2SSIP model has three characteristics: (i) The expected value function $\mathbb{E}\left[Q_{\boldsymbol{\xi}}(\mathbf{v}, \boldsymbol{\xi})\right]$ does not have a closed form and its values cannot be calculated easily; (ii) the optimal objective function expressed by Eq. (10.26) of the second-stage optimization problem can be calculated easily for a given first-stage decision and a realization of the random container shipment demand, by means of any efficient algorithm for solving linear programming problems; (iii) the number of feasible first-stage decisions is very large, meaning that enumeration approaches are not feasible. These three characteristics enable us to employ the sample average approximation (SAA) method proposed by Kleywegt, Shapiro, and Homem-De-Mello (2001) to solve the 2SSIP model.

The main procedure involved in using the SAA method to solve the 2SSIP model is as follows: First, a sample $\boldsymbol{\xi}_1, \cdots, \boldsymbol{\xi}_N$ of N realizations of the random container shipment demand vector $\boldsymbol{\xi}$ is generated, and the expected value function $\mathbb{E}\left[Q_{\boldsymbol{\xi}}(\mathbf{v}, \boldsymbol{\xi})\right]$ is approximated by the sample average function $N^{-1}\sum_{l=1}^{N} Q_{\boldsymbol{\xi}}(\mathbf{v}, \xi_l(\omega))$. The 2SSIP model, expressed by Eqs. (10.24)–(10.26), can thus be approximated by the SAA problem:

[SAA]

$$\min Z(\mathbf{v}) = \mathbf{c}^T\mathbf{v} + \frac{1}{N}\sum_{l=1}^{N} Q_{\boldsymbol{\xi}}(\mathbf{v}, \xi_l(\omega)) - \sum_{k\in\mathcal{K}} c_k^{\mathrm{OUT}} N_k^{\mathrm{MAX}}, \qquad (10.27)$$

subject to the constraints (10.10), (10.11), (10.13)–(10.16) and (10.25), where $Q_{\boldsymbol{\xi}}(\mathbf{v}, \xi_l(\omega))$ $(l = 1, 2, \ldots, N)$ is the optimal objective function value for the second-stage optimization problem with a given vector **v** and a given realization $\xi_l(\omega)$:

$$Q_{\boldsymbol{\xi}}(\mathbf{v}, \xi(\omega)) = \min \sum_{(o,\, d)\in\mathcal{W}} \sum_{h^{od}\in\mathcal{H}^{od}} \left(c^{h^{od}} - f^{od}\right) z^{h^{od}}\left(\xi^{od}(\omega)\right), \qquad (10.28)$$

subject to

$$\sum_{k\in\mathcal{K}} x_{kr}V_k \geq \sum_{(o,\,d)\in\mathcal{W}}\sum_{h^{od}\in\mathcal{H}^{od}} \rho_{ir}^{h^{od}} z_l^{h^{od}}\left(\xi_l^{od}(\omega)\right),\ \ \forall i=1,\cdots,m_r, r\in\mathcal{R} \tag{10.29}$$

$$\sum_{h^{(o,\,d)}\in\mathcal{H}^{(o,\,d)}} z_l^{h^{od}}\left(\xi_l^{od}(\omega)\right) \leq \xi_l^{od}(\omega),\ \ \forall(o,d)\in\mathcal{W} \tag{10.30}$$

$$z_l^{h^{od}}\left(\xi_l^{od}(\omega)\right) \geq 0,\ \ \forall(o,d)\in\mathcal{W},\ \ \forall h^{od}\in\mathcal{H}^{od}, \tag{10.31}$$

A Dual Decomposition and Lagrangian Relaxation Approach

It can be seen that the SAA problem expressed by Eqs. (10.27)–(10.31) involves N linear programming problems shown by Eqs. (10.28)–(10.31). Each of these linear programming problems corresponds to one realization (or scenario) of the random container shipment demand and needs to be solved to obtain the expected value associated with a given first-stage decision. One possible way to solve the SAA problem is as follows:

We first enumerate all feasible first-stage solutions, then calculate the value of the objective function shown in Eq. (10.27) with respect to each feasible first-stage decision, after solving the corresponding N linear programming problems. Finally, we choose a feasible first-stage decision with the minimum objective function value. This method might be workable for very small-scale problems, but not for real-life problems.

The dual decomposition and Lagrangian relaxation approach, proposed by Carøe and Schultz (1999), can be used to solve the SAA problem effectively because it can decompose the SAA problem into N subproblems based on the container shipment demand realization. To carry out the decomposition, the first-stage decision variables are duplicated with respect to each container shipment demand realization, as denoted by $\mathbf{v}_l, (l=1,2,\ldots,N)$. The SAA problem can be rewritten as follows:

$$Z_N = \min \frac{1}{N}\sum_{l=1}^{N} \mathbf{c}^T\mathbf{v}_l + \frac{1}{N}\sum_{l=1}^{N}\left[\sum_{(o,\,d)\in\mathcal{W}}\sum_{h^{od}\in\mathcal{H}^{od}}\left(c^{h^{od}} - f^{od}\right) z_l^{h^{od}}\left(\xi_l^{od}(\omega)\right)\right] - \sum_{k\in\mathcal{K}} c_k^{\text{OUT}} N_k^{\text{MAX}}, \tag{10.32}$$

subject to the constraints (10.10), (10.11), (10.13)–(10.16), (10.25), (10.29)–(10.31) and the nonanticipativity constraints (10.33):

$$\mathbf{v}_l = \cdots = \mathbf{v}_N, \tag{10.33}$$

The above nonanticipativity constraints imply that the first-stage decision should not depend on the container shipment demand realization that prevails in the second-stage optimization problem; they can be alternatively expressed as

$$\mathbf{v}_l = \mathbf{v}_{l+1}, l = 1, 2, \cdots, N-1. \tag{10.34}$$

Eqs. (10.34) can also be written using matrix notation:

$$\sum_{l=1}^{N} \mathbf{H}_l \mathbf{v}_l = \mathbf{0}, \tag{10.35}$$

where $\mathbf{H}_l$ $(l = 1, 2, \ldots, N)$ is a matrix with $(N-1) \times (2KR + K)$ rows and $2KR + K$ columns (i.e., $2KR + K$ is the total number of first-stage decision variables, n_{kr}^{TOTAL}, n_k^{IN} and x_{kr}), defined as follows:

$$\begin{array}{l} \mathbf{H}_1 = (\mathbf{I}, \mathbf{0}, \cdots, \mathbf{0})^T, \mathbf{H}_2 = (-\mathbf{I}, \mathbf{I}, \mathbf{0}, \cdots \mathbf{0})^T, \mathbf{H}_3 = (\mathbf{0}, -\mathbf{I}, \mathbf{I}, \cdots \mathbf{0})^T, \cdots, \\ \mathbf{H}_{N-1} = (\mathbf{0}, \cdots, -\mathbf{I}, \mathbf{I})^T, \mathbf{H}_N = (\mathbf{0}, \cdots \mathbf{0}, -\mathbf{I})^T \end{array} \tag{10.36}$$

where I and 0 are the square unity matrix and the zero matrix of size 2KR + K, respectively.

Let $\boldsymbol{\lambda}$ be a $(N-1) \times (2KR + K) -$ dimensional vector of Lagrangian multiplier associated with the nonanticipativity constraints shown in Eq. (10.35). The corresponding Lagrangian relaxation of the SAA problem can be formulated as shown below:

[LR]

$$LR(\boldsymbol{\lambda}) = \min \sum_{l=1}^{N} \left[\frac{1}{N} \mathbf{c}^T \mathbf{v}_l + \frac{1}{N} \sum_{(o,d) \in \mathcal{W}} \sum_{h^{od} \in \mathcal{H}^{od}} \left(c^{h^{od}} - f^{od} \right) z_l^{h^{od}} \left(\xi_l^{od}(\omega) \right) \right.$$
$$\left. + \boldsymbol{\lambda}^T \mathbf{H}_l \mathbf{v}_l - \frac{1}{N} \sum_{k \in \mathcal{K}} c_k^{\text{OUT}} N_k^{\text{MAX}} \right], \tag{10.37}$$

subject to the constraints (10.10), (10.11), (10.13)–(10.16), and (10.25), duplicated with respect to each container shipment demand realization, and the constraints (10.29)–(10.31) for each container shipment demand realization. This LR model can be further decomposed into N separate mixed-integer linear programming problems corresponding to the N container shipment demand realizations, namely

$$LR(\boldsymbol{\lambda}) = \sum_{l=1}^{N} LR_l(\boldsymbol{\lambda}), \tag{10.38}$$

where

$$\begin{aligned} LR_l(\boldsymbol{\lambda}) = \min \frac{1}{N}\mathbf{c}^T\mathbf{v}_l + \frac{1}{N}\sum_{(o,d)\in\mathcal{W}}\sum_{h^{od}\in\mathcal{H}^{od}}\left(c^{h^{od}} - f^{od}\right) z_l^{h^{od}}\left(\xi_l^{od}(\omega)\right) \\ + \boldsymbol{\lambda}^T\mathbf{H}_l\mathbf{v}_l - \frac{1}{N}\sum_{k\in\mathcal{K}} c_k^{\text{OUT}} N_k^{\text{MAX}}, \end{aligned} \tag{10.39}$$

subject to the constraints (10.10), (10.11), (10.13)–(10.16), (10.25), and (10.29)–(10.31) associated with the lth container shipment demand realization.

Each subproblem shown in Eq. (10.39) can be solved using an efficient optimization solver such as CPLEX. It is straightforward to demonstrate that $LR(\boldsymbol{\lambda})$, the optimal objective function value of the LR model with respect to a given Lagrangian multiplier $\boldsymbol{\lambda}$, is a lower bound on the optimal function value of the SAA problem (10.27). The best or tightest lower bound can be found by solving the Lagrangian dual model:

[LD]

$$LD = \max_{\boldsymbol{\lambda}} LR(\boldsymbol{\lambda}). \tag{10.40}$$

This Lagrangian dual model is a concave maximization problem with nondifferentiable objective function $LR(\lambda)$. Eqs. (10.37) and (10.38) further show that $\sum_{l=1}^{N}\mathbf{H}_l\mathbf{v}_l^*$ is a subgradient of the convex and nondifferentiable function $LR(\lambda)$, where $\mathbf{v}_l^*$ is the optimal solution of the lth subproblem shown in Eq.(10.39), namely:

$$\sum_{l=1}^{N}\mathbf{H}_l\mathbf{v}_l^* \in \partial LR(\lambda) \tag{10.41}$$

With this subgradient, the LR model can be solved using the following subgradient method:

Step 0: Take an initial Lagrangian multiplier vector $\boldsymbol{\lambda}^{(1)}$ and a predetermined step size sequence $\left\{t^{(1)}, t^{(2)}, \cdots\right\}$. Let the number of iterations $h = 1$.

Step 1: Calculate the subgradient $\sum_{l=1}^{N} \mathbf{H}_l \mathbf{v}_l^{*(h)}$ by solving the subproblem shown in Eq. (10.39), with respect to the Lagrangian multiplier vector $\boldsymbol{\lambda}^{(h)}$.

Step 2: Update the Lagrangian multiplier vector according to the formula:

$$\boldsymbol{\lambda}^{(h+1)} = \boldsymbol{\lambda}^{(h)} + t^h \sum_{l=1}^{N} \mathbf{H}_l \mathbf{v}_l^{*(h)}. \tag{10.42}$$

Step 3: If the following criterion is fulfilled, then the algorithm is terminated. Otherwise, let $h = h + 1$ and go to Step 1:

$$\left| \left(LR\left(\boldsymbol{\lambda}^{(h+1)}\right) - LR\left(\boldsymbol{\lambda}^{(h)}\right)\right) \Big/ LR\left(\boldsymbol{\lambda}^{(h)}\right) \right| \leq \tau, \tag{10.43}$$

where τ is a given tolerance.

The global convergence of this subgradient method has already been proven, provided that the step size satisfying the square is summable, but the step size conditions are not summable (see Shore, 1985):

$$t^{(h)} \geq 0 (h = 1, 2, \ldots, \infty), \sum_{h=1}^{\infty} t^{(h)} = \infty \text{ and} \sum_{h=1}^{\infty} \left[t^{(h)}\right]^2 < \infty \tag{10.44}$$

This study adopts the typical step size sequence $\left\{t^{(h)} = \frac{1}{h}, h = 1, 2, \cdots, \infty\right\}$ that fulfills the above condition.

Sample Average Approximation Method

The SAA method proposed by Kleywegt et al. (2001) is a Monte Carlo simulation-based approach to the 2SSIP model. The quality of the solution obtained from the SAA method, depending on the number of samples and the sample size, can be assessed by statistical analysis techniques. The SAA method, incorporating the dual decomposition and Lagrangian relaxation approach, as well as the relevant statistical moment estimation, is presented as follows:

Step 0: Generate M samples of the random container shipment demand, where each sample has size N, namely, $\{(\boldsymbol{\xi}_1^m, \cdots, \boldsymbol{\xi}_l^m, \cdots, \boldsymbol{\xi}_N^m) | m = 1, 2, \cdots, M\}$.

Step 1: Solve the SAA problem corresponding to each container shipment demand sample $(\boldsymbol{\xi}_1^m, \cdots, \boldsymbol{\xi}_l^m, \cdots, \boldsymbol{\xi}_N^m)$, $m = 1, 2, \ldots, M$, by using the above mentioned dual decomposition and Lagrangian relaxation approach. Let $\hat{\mathbf{v}}_N^m$ and $\hat{v}_N^m$ $(m = 1, 2, \ldots, M)$ be the relevant solution and the objective function value of the SAA problem (10.27), respectively.

Step 2: Calculate a point estimation of a lower bound on the optimal function value of the 2SSIP model (10.24) using

$$L_N^M = \frac{1}{M} \sum_{m=1}^{M} \hat{v}_N^m. \tag{10.45}$$

The variance of L_N^M can be estimated as follows:

$$\sigma_{L_N^M}^2 = \frac{1}{M(M-1)} \sum_{m=1}^{M} \left(\hat{v}_N^m - L_N^M\right)^2. \tag{10.46}$$

Step 3: For each optimal solution $\hat{\mathbf{v}}_N^m$ obtained in Step 1 $(m = 1, 2, \ldots, M)$, independently generate another container shipment demand sample $\left(\hat{\boldsymbol{\xi}}_1^m, \hat{\boldsymbol{\xi}}_2^m, \ldots, \hat{\boldsymbol{\xi}}_{\hat{N}}^m\right)$, where the sample size here is $\hat{N}$ ($\hat{N}$ is much larger than N), and calculate

$$\hat{v}_{\hat{N}}^m\left(\hat{\mathbf{v}}_N^m\right) = \mathbf{c}^T \hat{\mathbf{v}}_N^m + \frac{1}{\hat{N}} \sum_{l=1}^{\hat{N}} Q_{\boldsymbol{\xi}}\left(\hat{\mathbf{v}}_N^m, \hat{\boldsymbol{\xi}}_l\right) - \sum_{k \in \mathcal{K}} c_k^{\mathrm{OUT}} N_k^{\mathrm{MAX}}. \tag{10.47}$$

$\hat{v}_{\hat{N}}^m\left(\hat{\mathbf{v}}_N^m\right)$ is an unbiased estimation of an upper bound for the optimal function value of the 2SSIP model, because $\hat{\mathbf{v}}_N^m$ is one of its feasible solutions. The best upper bound is given by

$$U_{\hat{N}}^M = \min_{m \in \{1, \cdots, M\}} \left\{\hat{v}_{\hat{N}}^m\left(\hat{\mathbf{v}}_N^m\right)\right\}. \tag{10.48}$$

Let $\hat{\mathbf{v}}_N^{m^*} \in \arg\min\left\{\hat{v}_{\hat{N}}^m\left(\hat{\mathbf{v}}_N^m\right) | m = 1, \ldots, M\right\}$, so that the variance of $U_{\hat{N}}^M$ can be estimated by

$$\sigma_{U_{\hat{N}}^M}^2 = \frac{1}{\hat{N}(\hat{N}-1)} \sum_{l=1}^{\hat{N}} \left[\left(\mathbf{c}^T \hat{\mathbf{v}}_N^{m^*} + Q_{\boldsymbol{\xi}}\left(\hat{\mathbf{v}}_N^{m^*}, \hat{\boldsymbol{\xi}}_l\right) - \sum_{k \in \mathcal{K}} c_k^{\mathrm{OUT}} N_k^{\mathrm{MAX}}\right) - \hat{v}_{\hat{N}}^m\left(\hat{\mathbf{v}}_N^{m^*}\right)\right]^2. \tag{10.49}$$

Step 4: Calculate an estimate of the gap between L_N^M and $U_{\hat{N}}^M$ as follows:

$$\theta_{M,N,N'} = U_{N'}^M - L_N^M \tag{10.50}$$

The estimate of the variance of the gap estimator $\theta_{M,N,N'}$ is given by:

$$\sigma_{\theta_{M,N,N'}}^2 = \sigma_{U_{N'}^M}^2 + \sigma_{L_N^M}^2. \tag{10.51}$$

It has been proved by Norkin, Pflug, and Ruszczyński (1998) and Mak, Morton, and Wood (1999) that the expected value of $\hat{v}_N$ is less than or equal to the optimal value v^* of the original problem (10.24), namely $\mathbb{E}[\hat{v}_N] \leq v^*$. However, it is impossible to get the exact value of $\mathbb{E}[\hat{v}_N]$, which indicates that $\mathbb{E}[\hat{v}_N]$ has to be approximated by its sample mean, as the sample mean is an unbiased estimator of $\mathbb{E}[\hat{v}_N]$. In order to get the sample mean, M samples are generated (i.e., Step 0) and then the sample mean, denoted by L_N^M, is calculated (i.e., Step 1) using Eq. (10.45). As the calculated sample average L_N^M in Eq. (10.45) is an unbiased estimator of $\mathbb{E}[\hat{v}_N]$, L_N^M is less than or equal to the optimal value v^* of the original problem; thus L_N^M is a lower bound for v^*.

NUMERICAL EXAMPLE

In this section, we validate the applicability of the proposed methodology by implementing it on a numerical example shown in Table 8.1. We first do a sensitivity analysis of the SAA parameters, and finally explore the effect of the variance value of uncertain container shipment demand on the objective function value of the stochastic programming solution and comment on the quality of the stochastic programming solutions in comparison to those obtained using a deterministic approach. It is assumed that the short-term planning horizon in this numerical example is six months. The random container shipment demand between each O-D pair $(o, d) \in \mathcal{W}$ is assumed to follow the normal distribution; that is, $\xi^{od} \sim N(\mu^{od}, \sigma^{od})$ with truncated tails, which is the sample space that comprises positive values within a given range. The rationale behind using the normal distribution is that the deviation between forecasted and real demand is often approximately normally distributed (Brown, 1959). For the sake of presentation the ratio μ^{od}/σ^{od} is assumed to be the same for all O-D port pairs.

We set the stop tolerance $\tau = 10^{-6}$ in the subgradient method, as well as the number of samples $M = 20$ and $\hat{N} = 1000$ in the SAA method.

The solution algorithm is programmed using the programming language Lua (v5.1) with a mixed-integer linear programming solver. All computations are carried out on a desktop personal computer with Intel Core 2 CPU 1.86 GHz and 2.0 GB of RAM under Microsoft Windows 7.

Sensitivity Analysis of the Sample Size *N* in the SAA Method

Table 10.2 gives the lower bound, upper bound, gap, and 95% confidence interval of the gap for each sample size $N \in \{20, 30, 40, 50, 60\}$, which was obtained using the proposed solution method. According to Table 10.2, the confidence interval of the optimality gap becomes narrower as the sample size increases. We thus take the sample size $N = 60$ in the subsequent analysis, in view of the acceptable confidence interval that results from this sample size.

Results Discussions

We now investigate the effect of container shipment demand uncertainty by comparing the average profits obtained from the proposed 2SSIP model to those obtained from the expected value problem (EVP) that is, the profits obtained where the uncertain container shipment demands are replaced by their mean values from the 2SSIP model. After solving the EVP the optimal first-stage solutions (i.e., the fleet design and deployment decisions) are obtained. Given these optimal first-stage solutions obtained from the EVP, we compute the EEV (see Birge & Louveaux, 1997), which i the expected value of the EVP solution, by computing the expected value of the EVP first-stage solution across a large number of different scenarios of container shipment demand.

In order to investigate the effect of container shipment demand uncertainty, three different levels (low, medium, and high) of standard deviation

Table 10.2 Statistical Lower Bound, Upper Bound, Estimated Gap and Confidence Interval With $M = 20$ and $\hat{N} = 1000$

	Lower Bound ($\times 10^6$)		Upper Bound ($\times 10^6$)		Estimated Gap ($\times 10^6$)		95% Confidence Interval ($\times 10^6$)		
N	**Aver**	σ_{LB}	**Aver**	σ_{UB}	**Aver**	σ_{gap}	**Min**	**Max**	**Interval**
20	8875.5	1.9	8877.8	1.1	2.3	2.2	−1.2	5.9	7.2
30	8876.9	1.8	8878.5	1.1	1.6	2.1	−1.7	5.1	6.8
40	8875.4	1.4	8877.6	1.2	2.2	1.9	−0.6	5.2	5.8
50	8875.2	1.3	8877.1	1.1	1.9	1.7	−0.8	4.8	5.7
60	8874.7	0.9	8877.1	1.2	2.3	1.4	−0.01	4.6	4.6

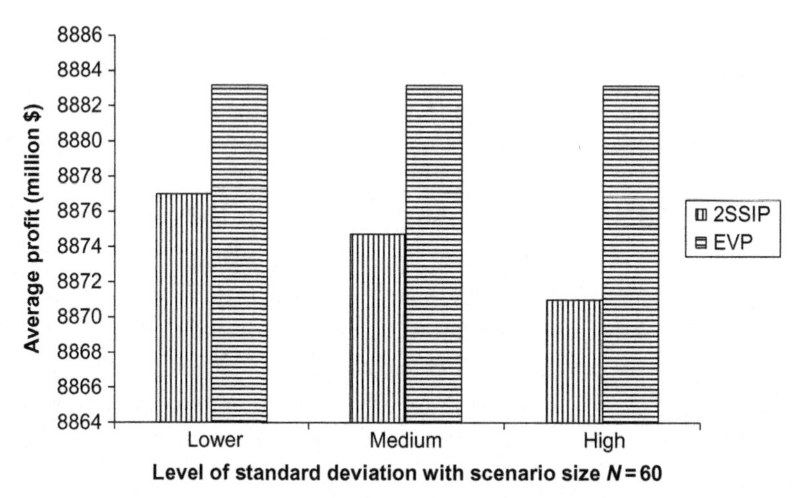

Fig. 10.2 Average profits of 2SSIP model and EVP with scenario size $N=60$.

of demand are considered: 5%, 10%, and 15% of the mean value of demand, respectively. In Fig. 10.2, the average profits obtained using the 2SSIP model, corresponding to these three levels, are compared with those obtained using the EVP. It is clearly observed that the estimated average profits in the 2SSIP model corresponding to all three levels of variance are smaller than those in the EVP. This is reasonable because in the EVP, the container shipment demands are deterministic rather than random, with their values given by the mean values used in the 2SSIP model; thus the EVP could be regarded as a problem with deterministic container shipment demand. Therefore the EVP with deterministic container shipment demand would be expected to have a higher yield than the 2SSIP model with uncertain container shipment demand. This indicates that the precision of the estimate of container shipment demand is significant for a liner container shipping company. Moreover, it is found that the expected profits decrease with an increase in the variance of container shipment demands. This further verifies the significance of container shipment demand information for the liner container shipping company.

After solving the EVP, the optimal first-stage solutions are obtained; these are the decisions about the numbers and types of ships in the fleet and the ship-to-route allocation. Then the EEV can be computed by implementing the EVP first-stage solution for a large number of different scenarios of container shipment demand. The EEVs and the expected profits from the 2SSIP model associated with low, medium, and high standard deviations of container shipment demand are depicted in Fig. 10.3. It is clear that

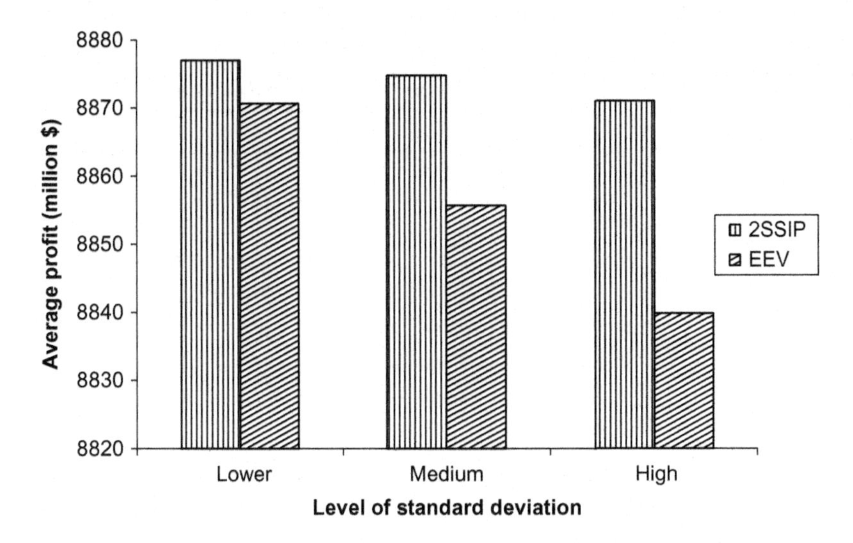

Fig. 10.3 Average profits of 2SSIP model and EEV for different level of variance.

the estimated average profits of the 2SSIP model corresponding to the three levels of variance are all higher than the EEVs, which shows that the 2SSIP model would be expected to have a higher yield than the EVP, and indicates that the 2SSIP model is superior to the EVP. Also, we find that the ratios between the objectives for 2SSIP and EEVs grow with increasing variance as expected. However, we have to acknowledge that the average profit obtained from the 2SSIP model is weak, because we can only set proper but not precise values of the SAA parameters, M, N, and $\hat{N}$. Additionally, although Shore (1985) proved that, theoretically, $LR(\boldsymbol{\lambda}^h) \rightarrow LD$ in the dual decomposition method, it is quite difficult to reach the convergence point in practice. We can only set a tolerance τ in order to find a relative better solution with an acceptable level of precision.

The liner ship fleet plans suggested by the 2SSIP model and the EVP with a low variance of container shipment demand are shown in Tables 10.3 and 10.4, respectively. Both fleets contain a total of 34 ships. However, the two plans are different. As can be seen from the results shown in Table 10.3, the fleet of the liner container shipping company consists of four types of ships, three ships are chartered out and ten ships are chartered in; while in Table 10.4, its fleet contains five types of ship, eight ships are chartered out and one ship is chartered in.

Table 10.3 Liner Ship Fleet Plan Produced by the 2SSIP Model With Low Variance

	Route Type							
Ship Type	**CCX**	**CPX**	**GIS**	**IDX**	**NCE**	**NZX**	**SCE**	**UKX**
Ship allocations								
1			2			2	3	
2								1
3	3	3	1			1	3	
4								
5				5	8		2	
Number of voyages								
1			15			17	8	
2								26
3	28	28	12			10	12	
4								
5				32	31		8	

Table 10.4 Liner Ship Fleet Plan Produced by the EVP

	Route Type							
Ship Type	**CCX**	**CPX**	**GIS**	**IDX**	**NCE**	**NZX**	**SCE**	**UKX**
Ship allocations								
1			2			2	3	
2							1	1
3	3	3	1			1	2	
4				1				
5				4	8		2	
Number of voyages								
1			15			16	11	
2							4	26
3	27	27	11			10	7	
4				6				
5				28	31		8	

SUMMARY

This paper has studied a realistic planning problem with container transshipment and demand uncertainty that a liner container shipping company has experienced. The problem was formulated as a 2SSIP model, but it is actually possible to adapt the mathematical formulation of the problem to any planning problem that consists of two stages of decision variables. The greatest difficulty in solving the 2SSIP model is determining how to deal with the expected recourse function, which is only implicitly defined, depends on the first-stage decisions, and usually involves optimization problems embedded in expectation. To effectively solve the proposed model, we first use the sample average approximation method to approximate the expected recourse function. Next the dual decomposition and Lagrangian relaxation method are used to solve the model. The proposed model and solution methods are tested using a numerical example. The gaps between the lower and upper bounds are small, which indicates that the solution methods are effective. It is also found that the variability of the uncertain parameters has a significant effect on the solutions. There are three future research directions we will pursue. First, in this paper, we assume that only bareboat charter is adopted. It is worthwhile to investigate the other two types of chartering ships in future studies. We will also investigate the long-term planning problem in the future. In the long-term LSFP problem, another business behavior needs to be included, namely purchasing new ships. The formulation and solution method could thus be more complicated. A third research topic would be to examine the operational-level decision problems; however, more detailed characteristics of liner shipping must be captured to address operational problems.

REFERENCES

Birge, J. R., & Louveaux, F. V. (1997). *Introduction to stochastic programming*. New York, NY: Springer-Verlag.

Brown, R. G. (1959). *Statistical forecasting for inventory control*. New York, NY: McGraw-Hill.

Carøe, C. C., & Schultz, R. (1999). Dual decomposition in stochastic integer programming. *Operations Research Letters*, *24*, 37–45.

Christiansen, M., Fagerholt, K., & Ronen, D. (2004). Ship routing and scheduling: status and perspectives. *Transportation Science*, *38*(1), 1–18.

Cullinane, K., & Khanna, M. (1999). Economies of scale in large container ships. *Journal of Transport Economics and Policy*, *33*(2), 185–207.

Kleywegt, A. J., Shapiro, A., & Homem-De-Mello, T. (2001). The sample average approximation method for stochastic discrete optimization. *SIAM Journal on Optimization*, *12*, 479–502.

Mak, W. K., Morton, D. P., & Wood, R. K. (1999). Monte carlo bounding techniques for determining solution quality in stochastic programs. *Operations Research Letters*, *24*(1), 47–56.

Mourão, M. C., Pato, M. V., & Paixão, A. C. (2001). Ship assignment with hub and spoke constraints. *Maritime Policy and Management*, *29*(2), 135–150.

Norkin, V. I., Pflug, G. C., & Ruszczyński, A. (1998). A branch and bound method for stochastic global optimization. *Mathematical Programming*, *83*, 425–450.

Perakis, A. N. (2002). Fleet operation optimization and fleet deployment. In Th. Costas & Th. Grammernos (Eds.), *The handbook of maritime economics and business* (pp. 580–597). London, UK: Lloyds of London Publications.

Ronen, D. (1983). Container ships routing and scheduling: Survey of models and problems. *European Journal of Operational Research, 12*(2), 119–126.

Ronen, D. (1993). Ship scheduling: The last decade. *European Journal of Operational Research, 71*(3), 325–333.

Shore, N. Z. (1985). *Minimization methods for non-differentiable functions*. New York, NY: Springer-Verlag Inc.

Vernimmen, B., Dullaert, W., & Engelen, S. (2007). Schedule unreliability in liner shipping: Origins and consequences for the hinterland supply chain. *Maritime Economics and Logisitics, 9*, 193–213.

CHAPTER ELEVEN

Multiperiod Liner Ship Fleet Planning

Contents

INTRODUCTION

This chapter studies a realistic long-term (multiperiod) LSFP problem with container transshipment for a liner container shipping company. Traditional multiperiod liner ship fleet planning begins with a forecasted container shipment demand of each single period based on forecasting techniques, such as regression and time series models. However the forecasted container shipment demand, which is a key input in a multiperiod LSFP problem, can never be forecasted with complete confidence. Moreover,

Liner Ship Fleet Planning
http://dx.doi.org/10.1016/B978-0-12-811502-2.00011-9

the historical data fully show that the current container shipment demand has an effect on the future demand, which indicates the container shipment demands of different periods are dependent. Therefore, it is realistic and necessary to take the uncertainty and dependency of container shipment demand into account in a multiperiod LSFP problem. Here, the container shipment demand between two successive single periods is assumed to be dependent. During a multiperiod planning horizon, the container shipment demand in each single period is possibly different, which implies that the liner ship fleet plans vary with periods and depend on container shipment demand. Therefore to cope with the period-dependent container shipment demand pattern, the liner container shipping company has to adjust each period its liner ship fleet plan of determining fleet size, mix, and deployment.

Under the consideration of both container transshipment and the uncertainty and dependency of container shipment demand, the classical multiperiod LSFP problem, which is studied under a deterministic environment and without container transshipment, could be expanded into a fresh and worthwhile research area. This chapter thus focuses on model development and the design of solution methods for the multiperiod LSFP problem with container transshipment, as well as uncertain and dependent container shipment demand.

Multiperiod/long-term ship fleet planning problems have been studied for several decades. However, the existing researches on the topic all make the assumption that the demand is known and deterministic. Nicholson and Pullen (1971) were pioneers in the field, having developed a dynamic programming model for a ship fleet management problem that aimed to find the best sale and replacement policy with the objective of maximizing the multiperiod company assets. They proposed a two-stage decision strategy: The first stage determines a priority order for selling a ship, based on its assessment of the net contribution to the objective function if it is sold each year, regardless of the rate at which charter ships are taken on. The second stage uses the dynamic programming approach to find the optimal level of chartering for a given priority replacement order. Cho and Perakis (1996) developed an integer linear programming model for a multiperiod liner ship fleet planning problem seeking to determine the optimal fleet size, mix, and ship-to-route allocation. In their model, as long as those decisions are made at the beginning of the planning horizon, they remain static over the whole horizon. Such a period-independent model cannot characterize a realistic dynamic decision strategy; the fleet size, mix, and ship-to-route allocation should be adjustable period by period, since the container shipment demand is period dependent. In other words, it is more rational and practical to assume that the fleet size, mix, and ship-to-route allocation are period-dependent (dynamic) decisions rather than static ones.

Xie, Wang, and Chen (2000) thus reformulated the multiperiod liner shipping problem proposed by Cho and Perakis (1996) by applying a dynamic programming approach. They first divided the multiperiod planning horizon into a number of single periods, with each period being one year. For each period, they used integer linear programming to determine the fleet size, mix, and ship-to-route assignment, incurring minimal cost. However, they assumed that the annual operating cost and transportation capacity of each ship on each route were constant. This assumption is unrealistic because the costs are voyage-dependent. For example, a ship sailing 20 voyages on a given route over a given year would certainly incur greater annual operating costs and have a greater transportation capacity than a ship that sails 10 voyages on the same route. Recently, Meng and Wang (2011) proposed a realistic multiperiod LSFP problem for a liner container shipping company and formulated this problem as a scenario-based dynamic programming model. However, in addition to the deterministic container shipment demand assumption, these studies (i.e., Cho & Perakis, 1996; Meng & Wang, 2011; Nicholson & Pullen, 1971; Xie et al., 2000) do not take container transshipment operations into account.

When compared with the few relevant papers on the MPLSFP problem with uncertain container shipment demand, much research has been devoted to other problems under the assumption of uncertain multiperiod demand, such as capacity expansion problems (Ahmed, King, & Parija, 2003; Ahmed & Sahinidis, 2003; Berman, Ganz, & Wagner, 1994; Laguna, 1998; Singh, Philpott, & Wood, 2009; Wagner & Berman, 1995), airline fleet composition and allocation problems (Listes & Dekker, 2005), a multisite production planning problem (Leung, Lai, Ng, & Wu, 2007; Leung, Tsang, Ng, & Wu, 2007), portfolio management problems (Celikyurt & Özekici, 2007; Gülpinar & Rustem, 2007; Osorio, Gülpinar, & Rustem, 2008), and others. Their objectives are to minimize or maximize the expected value of a key variable, such as cost or profit, over a multiperiod planning horizon, which is defined as the sum of the cost or profit in each single period. However, the methodologies applied or proposed in those studies did not involve the dependency of the uncertain multiperiod demand. Shapiro and Philpott (2007) did, in fact, previously mention the dependency of uncertain demand in a multistage stochastic programming problem. Unfortunately, no application or study involving dependency has been reported so far.

Therefore the model formulation for the multiperiod LSFP problem integrating uncertainty and dependency is a challenge and a goal of this chapter. The objective is to seek an optimal multiperiod liner ship fleet plan (i.e., a joint ship fleet development and deployment plan for the multiperiod planning horizon) that will be implemented before the container shipment demands are known, such that the expected profit reaped across the whole multiperiod planning horizon is maximized by a liner container shipping company implementing this plan.

To formulate the uncertainty of container shipment demand during a particular period within the multiperiod planning horizon, it is assumed to be a discrete random variable taking a limited number of possible values with a known occurrence probability. It has to be pointed out that the container shipment demand in a given period is dependent on the demand in previous periods. Therefore the probability is actually a conditional probability so as to reflect this dependency.

In order to capture a characteristic of the realistic dynamic planning strategy, the multiperiod planning horizon is divided into a number of single periods, and a stochastic programming model is developed for each one, with the aim of determining fleet deployment for that period. Using a scenario tree approach to model the evolution of the dependent uncertain demand of two successive single periods, as well as a decision tree to interpret the procedure of liner ship fleet planning, the proposed problem in this study is formulated as a multiperiod stochastic programming model. This type of model comprises a sequence of interrelated stochastic programming models developed for each single period. We further show that the multiperiod stochastic programming model can be equivalently transformed into a shortest-path problem defined on an acyclic network. A path on the acyclic network corresponds to a multiperiod liner ship fleet planning. Finally, a numerical example is carried out to assess the applicability and performance of the proposed model and solution algorithm.

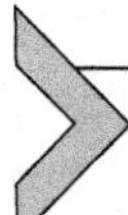

PROBLEM STATEMENT

Uncertainty and Dependency of Container Shipment Demand

Assume that the multiperiod planning horizon consists of T single periods, denoted by $\mathcal{T} = \{1, \ldots, t, \ldots, T\}$. The length of one single period can be determined according to the changes in container shipment demand forecasted within the multiperiod planning horizon; for example, one period could be one year. Let ξ_t^{od} be the number of containers in terms of TEUs (20-foot equivalent unit) to be transported between an O-D port pair $(o, d) \in \mathcal{W}$ in a particular single period $t \in \mathcal{T}$. The uncertainty of container shipment demand in the multiperiod LSFP problem is included in the model by specifying a set of discrete demand scenarios. Let $\mathcal{S}_t = \{1, \ldots, s \ldots, S_t\}$ be the set of container shipment demand scenarios for period $t \in \mathcal{T}$. In each scenario $s \in \mathcal{S}_t$, values are specified for the container shipment demand between each port pair in period $t \in \mathcal{T}$. $s \in \mathcal{S}_t$ is a weight that is associated with each scenario. These weights are often thought of as the probabilities that each scenario will occur; they are denoted by p_s^t and characterized by $\sum_{s=1}^{S_t} p_s^t = 1$. In other words the container shipment demand between each port pair during a particular period, namely $\xi_t^{od}((o, d) \in \mathcal{W}, t \in \mathcal{T})$, is assumed to be a discrete random variable, which takes a limited number of possible values with known occurrence probabilities.

Moreover, the historical data fully show that the current container shipment demand has an effect on future demand, which indicates that the container shipment demand in one period is dependent on that of previous periods. As the effect on demand in some faraway future is quite weak, we simply assume that the container shipment demand is only dependent on that of the previous period. Therefore the scenario $s \in \mathcal{S}_t$ is dependent on the scenario $s' \in \mathcal{S}_{t-1}$. Let $p_{s|s'}^t$ be the conditional probability that scenario s occurs in period t; given that scenario s' occurs in period t–1, then p_s^t is given by $\sum_{s'=1}^{S_{t-1}} p_{s'}^{t-1} p_{s|s'}^t$.

As scenario $s \in \mathcal{S}_t$ occurs in period t with conditional probability $p_{s|s'}^t$, given that scenario $s' \in \mathcal{S}_{t-1}$ occurs in period $t-1$, all scenarios for the whole T-period planning horizon can be depicted as a scenario tree with T layers, where each layer corresponds to a single period. The following example is provided for clarity:

An Example

Let us consider the liner shipping service route shown in Fig. 11.1 in order to illustrate the scenarios of container shipment demand. For simplicity, consider two periods (e.g., two years) and three O-D pairs: Pusan (PS) → Shanghai (SH), Shanghai (SH) → Yantian (YT), Yantian (YT) → Hong Kong (HK). Suppose that there are three discrete scenarios of demand in each year: L (low), M (medium), and H (high), as shown in Table 11.1. These six scenarios for the two years are illustrated by a two-layer scenario tree, shown in Fig. 11.2.

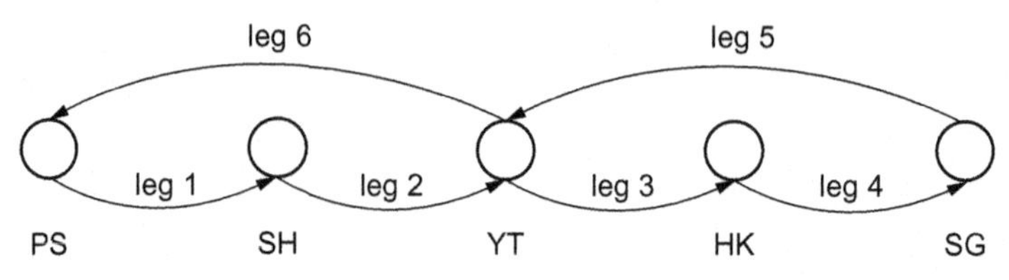

Fig. 11.1 A liner shipping service route.

Table 11.1 Container Shipment Demand Scenarios for Illustrative Example

	Year 1			Year 2		
O-D Pairs	**Low**	**Medium**	**High**	**Low**	**Medium**	**High**
PS → SH	1000	2000	3000	1500	2500	3500
SH → YT	800	1000	1500	1200	2000	2500
YT → HK	1000	1500	2000	1500	2000	2500

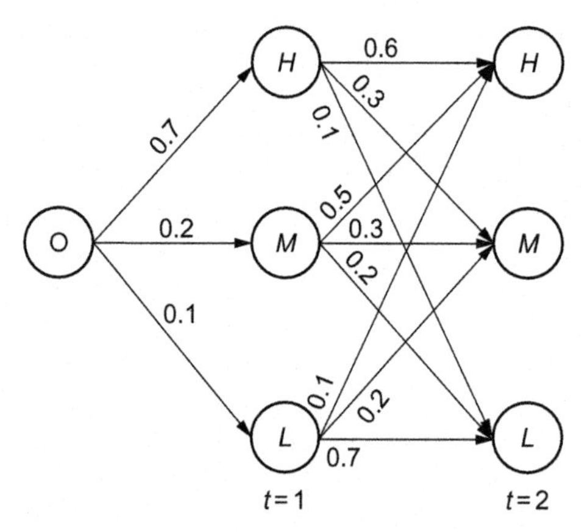

Fig. 11.2 Probability of scenarios for the illustrative example.

The value on each branch in the two-layer scenario tree denotes the probability or conditional probability of each scenario's occurrence. Accordingly the probabilities of each of the three scenarios in year 2 are computed as follows:

$$
\begin{aligned}
p_H^2 &= p_H^1 \times p_{H|H}^2 + p_M^1 \times p_{H|M}^2 + p_L^1 \times p_{H|L}^2 = 0.7 \times 0.6 + 0.2 \times 0.5 + 0.1 \times 0.1 = 0.53 \\
p_M^2 &= p_H^1 \times p_{M|H}^2 + p_M^1 \times p_{M|M}^2 + p_L^1 \times p_{M|L}^2 = 0.7 \times 0.3 + 0.2 \times 0.3 + 0.1 \times 0.2 = 0.29 \\
p_L^2 &= p_H^1 \times p_{L|H}^2 + p_M^1 \times p_{L|M}^2 + p_L^1 \times p_{L|L}^2 = 0.7 \times 0.1 + 0.2 \times 0.2 + 0.1 \times 0.7 = 0.18
\end{aligned}
\tag{11.1}
$$

Fleet Size and Mix Strategies

The liner container shipping company can use its own ships to pick up and deliver containers for shippers; it may also charter ships from other liner container shipping companies or purchase new ships to meet its container shipment demand. The company may also charter out some of its own ships, depending on their capacity in terms of TEUs. A fleet size and mix strategy associated with a particular period within the T-period planning horizon is defined as a plan comprising the number of ships to be chartered, the number of the company's own ships to be chartered out, the number of its own ships to be used during the period, and the number of new ships to be purchased. The order time for new ships is ignored, as this is generally known in practice.

At the beginning of the period $t \in \mathcal{T}$, experts from the strategic development department of the liner container shipping company would propose several possible fleet size and mix strategies for the period, based on their experiences and/or the available budget of the company for the period. It is thus assumed that there is a number of suggested fleet size and mix scenarios at the beginning of each period $t \in \mathcal{T}$. There is an inherent and implicit relation between these strategies from one period to the next. For example, assuming that the liner container shipping company currently owns three ships named by A, B, and C, the experts might propose two possible fleet size and mix strategies at the beginning of period t. Strategy 1 might be to use the existing three ships, while strategy 2 might be to purchase a new ship D to use as well. The two strategies would lead to two different states of the ship fleet at the beginning of the next period $t+1$: In the first state, there are three ships in the fleet, while in the second state there are four. Each of these two states becomes a possible initial state of the fleet at the beginning of period $t+1$. At the beginning of period $t+1$ the experts will propose a group of possible fleet size and mix strategies with respect to each of the two ship fleet states. This strategy decision process will be repeated until the end of the last period T; that is, the beginning of period $T+1$. Thus the entire decision process of fleet size and mix strategies actually forms a decision tree containing T layers.

Multiperiod Liner Ship Fleet Planning

The multiperiod LSFP problem with container transshipment and uncertain container shipment demand aims to maximize the total expected profit reaped over the whole T-period planning horizon by making an optimal joint fleet development and deployment plan. A joint fleet development and deployment plan consists of (1) a fleet size and mix strategy proposed by the experts at the beginning of each period (i.e., a fleet development plan), and (2) a fleet deployment plan, including the allocation of the ships in the fleet-to-liner ship routes; that is, the number of voyages by each ship on each liner shipping route $r \in \mathcal{R}$ required to maintain a given liner shipping service frequency on the route and the number of lay-up days allocated to each ship for maintenance. The objective of the deployment plan is expected profit maximization under various scenarios of container shipment demand for each of the given fleet size and mix strategies.

The rationale behind the adoption of this period-by-period planning is that the liner container shipping company can be flexible in adjusting its ship fleet size and mix according to the varying container shipment demand in each period. Moreover, the ships are assets with finite lives, which implies that the ship fleet needs to be renewed as old ships are removed from the fleet when they reach a given age limit and new ships are added in their place. The adoption of period-by-period planning thus also satisfies the physical requirement of the renewal of the fleet over time. We assume the liner container shipping company makes its planning decisions at the beginning of each single period, and this process is repeated until all periods in the multiperiod planning horizon have been covered. Therefore the multiperiod fleet plan consists of a number of single-period fleet plans. At the end of the planning horizon, we assume that all ships owned by the liner container shipping company are disposed of for their salvage values.

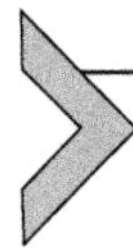

MODEL DEVELOPMENT

Decision Tree of Fleet Development Plan

The procedure of determining a fleet development plan for a T-period planning horizon can be interpreted as a decision tree with T layers, where each layer represents a period, as well as where each node in layer t of the tree represents a fleet size and mix strategy proposed at the beginning of period t. Dummy node O is introduced as the root of the decision tree to represent the current ship fleet state; that is, the decision tree grows from the root O. Each node in period t $(t = 1, 2, \ldots, T-1)$ can be regarded as a parent and will generate some offspring in period $t+1$; that is, the fleet size and mix strategies for the next period. Each parent and its offspring are connected by an arc, and it is noted that different parents may produce the same offspring. Each node of the decision tree, except the root, has a parent (which may not be unique). A parent n at period t and its offspring from period $t=1,\ldots,T-1$ to the end of the whole T-period planning horizon form a subtree, denoted by $\mathbb{T}^t(n)$. Each parent n, namely a nonterminal node in period $t=1,\ldots,T-1$, is the root of the sub-tree $\mathbb{T}^t(n)$. Thus $\mathbb{T}^0$ denotes the entire tree over the whole T-period planning horizon. The set of paths from root O to a node n in period t, is denoted by $\mathbb{P}^t(n)$, and each path $l \in \mathbb{P}^t(n)$ represents a development plan of fleet sizes and mixes for t periods. If n is a terminal node (i.e., a leaf), then path l corresponds to a development plan for all T periods.

Fig. 11.3 schematically illustrates the decision tree. In this figure, let $\mathcal{N}_t=\{1, \ldots, N_t\}$ be the set of nodes in period $t \in \mathcal{T}$, where N_t is the number of nodes in this set. Let $\mathcal{N}_t^m = \{1, \ldots, N_t^m\}$ be the set of strategies proposed for period $t+1$, which are generated from a particular strategy m proposed

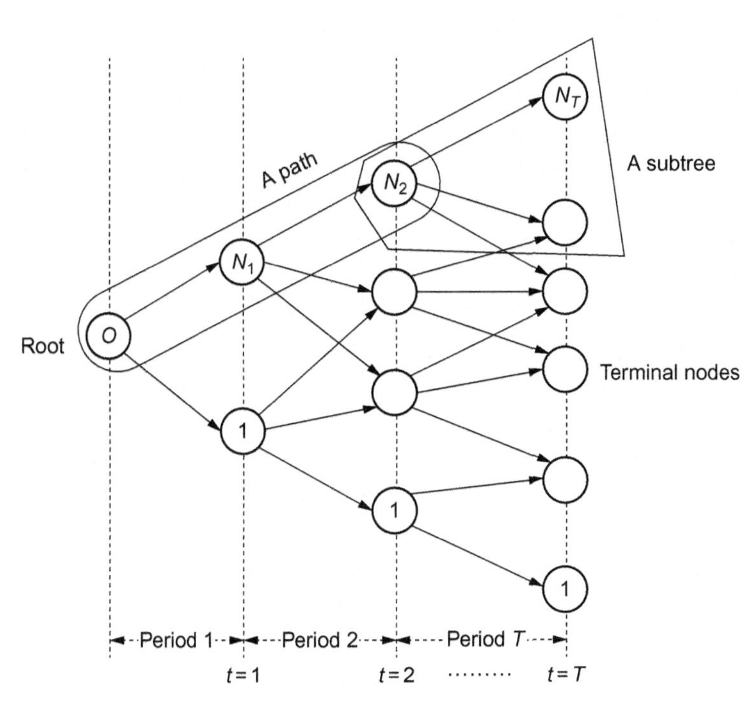

Fig. 11.3 Decision tree for fleet development plan.

for period t, where N_t^m represents the number of strategies of the set $\mathcal{N}_t^m$. If each offspring node has a unique parent, we then have:

$$\mathcal{N}_{t+1} = \bigcup_{m \in \mathcal{N}_t} \mathcal{N}_t^m, \quad t = 1, \ldots, T-1 \tag{11.2}$$

and

$$N_{t+1} = \sum_{m=1}^{N_t} N_t^m, \quad t = 1, \ldots, T-1 \tag{11.3}$$

The following notations are introduced for the sake of presentation:

$\mathcal{G}_{t,n}^{\text{KEEP}}$, set of company's own ships to be used at the beginning of period t in strategy n

$\mathcal{G}_{t,n}^{\text{SOLD}}$, set of company's own ships to be sold at the beginning of period t in strategy n

$\mathcal{G}_{t,n}^{\text{OUT}}$, set of own ships to be chartered out at the beginning of period t in strategy n

$\mathcal{G}_{t,n}^{\text{IN}}$, set of ships to be chartered in at the beginning of period t in strategy n

$\mathcal{G}_{t,n}^{\text{NEW}}$, set of new ships bought at the beginning of period t in strategy n

$\mathcal{G}_{t,n}$, set of ships that are used to deliver containers at the beginning of period t in strategy n.

For a node (strategy) n in period t, the ships that can be used to deliver containers include the company's own ships, which are kept in service; new ships purchased at the beginning of period t ($\mathcal{G}_{t,n}^{\text{NEW}} = \varnothing$ if no available new ships); and ships chartered in from other liner container shipping companies. The set of ships used in strategy n to deliver containers is given by:

$$\mathcal{G}_{t,n} = \mathcal{G}_{t,n}^{\text{KEEP}} \cup \mathcal{G}_{t,n}^{\text{NEW}} \cup \mathcal{G}_{t,n}^{\text{IN}}, \quad \forall t \in \mathcal{T} \tag{11.4}$$

The relationship between a parent m in period t and its offspring n in period $t+1$ ($t = 1, \ldots, T-1$) is given by:

$$\begin{aligned} &\mathcal{G}_{t,m}^{\text{KEEP}} \cup \mathcal{G}_{t,m}^{\text{OUT}} \cup \mathcal{G}_{t,m}^{\text{NEW}} = \mathcal{G}_{t+1,n}^{\text{KEEP}} \cup \mathcal{G}_{t+1,n}^{\text{OUT}} \cup \mathcal{G}_{t+1,n}^{\text{SOLD}}, \\ &m = 1, \ldots, N_t, n = 1, \ldots, N_t^m, t = 1, \ldots, T-1 \end{aligned} \tag{11.5}$$

2SSP Models for Fleet Deployment Plans

Each node n in period $t \in \mathcal{T}$ represents a fleet size and mix strategy proposed by the liner container shipping company's experts, based on their experience and available budget, which is used for investment in the chartering in or purchase of new ships. However, the decisions of how to properly deploy the ships in the fleet, as given by the fleet size and mix strategy n in period $t \in \mathcal{T}$, in order to maximize the profit gained from shipping containers over period t, have not yet been determined. Four types of decision variables are now defined as follows:

δ_{nt}^{kr}, binary variables equal to 1, if ship k is assigned to route r in strategy n of period t and 0 otherwise

x_{nt}^{kr}, number of voyages sailed by ship k on route r in strategy n of period t

y_{nt}^{k}, number of lay-up days of ship k in strategy n of period t

$z_{snt}^{h^{od}}$, number of containers carried by ships deployed on the container route $h^{od} \in \mathcal{H}^{od}$ between O-D port pair $(o, d) \in \mathcal{W}$ under container shipment demand scenario s in strategy n of period t.

Given the set of ships under strategy n of period t, namely $\mathcal{G}_{t,n}$; the values of ξ_t^{od} for a port pair $(o, d) \in \mathcal{W}$ under scenario $s \in \mathcal{S}_t$ in period $t \in \mathcal{T}$, denoted by ω_{st}^{od}; and the freight rate of transporting a container from its origin port o to its destination port d in period t (\$/TEU), denoted by f_t^{od}, the revenue gained from shipping containers along all possible routes in period t under container shipment demand scenario s is given by:

$$\sum_{(o,d)\in\mathcal{W}} \sum_{h^{od}\in\mathcal{H}^{od}} f_t^{od} z_{snt}^{h^{od}}\left(\omega_{st}^{od}\right) \tag{11.6}$$

Other revenue gained in strategy n over period t includes earnings from chartering out the company's ships and the salvage value gained from selling its ships. This is given by the following:

$$\sum_{k\in\mathcal{G}_{t,n}^{\text{OUT}}} c_{kt}^{\text{OUT}} + \sum_{k\in\mathcal{G}_{t,n}^{\text{SOLD}}} c_{kt}^{\text{SOLD}} \tag{11.7}$$

where c_{kt}^{OUT} is the amount received for chartering out a particular ship k at the beginning of period t (\$) and c_{kt}^{SOLD} is the amount received for selling out a ship k at the beginning of period t (\$).

The total costs incurred in strategy n of period t usually consist of the following components: container handling costs, voyage costs of ships in the fleet to transport containers, lay-up costs of ships for maintenance, costs of chartering in ships from other liner container shipping companies, and capital investment of purchasing new ships. The container handling cost incurred in a container route includes container loading cost at origin port, container discharging cost at destination port, and container transshipment cost at transshipment ports, if any. Let $c_t^{h^{od}}$ (\$/TEU) denote the container handling cost per TEU incurred in the container route $h^{od} \in \mathcal{H}^{od}$ over period t and then the total container handling cost can be calculate by

$$\sum_{(o,d)\in\mathcal{W}}\sum_{h^{od}\in\mathcal{H}^{od}} c_t^{h^{od}} z_{snt}^{h^{od}}\left(\omega_{st}^{od}\right) \tag{11.8}$$

The voyage costs of the ships in the fleet that are used to transport containers, plus lay-up costs of those ships undergoing maintenance, the costs of chartering in ships from other liner container shipping companies, and the capital investment of purchasing new ships are given by:

$$\sum_{r\in\mathcal{R}}\sum_{k\in\mathcal{G}_{t,n}} c_{krt}x_{nt}^{kr} + \sum_{k\in\mathcal{G}_{t,n}} e_{kt}y_{nt}^{k} + \sum_{k\in\mathcal{G}_{t,n}^{\text{IN}}} c_{kt}^{\text{IN}} + \sum_{k\in\mathcal{G}_{t,n}^{\text{NEW}}} c_{kt}^{\text{NEW}} \tag{11.9}$$

where c_{krt} is the voyage cost of operating a specific ship k on route r in period t (\$/voyage), e_{kt} is the daily lay-up cost for a specific ship k in period t (\$/day), c_{kt}^{IN} is the cost of chartering in a specific ship k at the beginning of period t (\$), and c_{kt}^{NEW} is the price of the new ship k at the beginning of period t (\$).

As mentioned earlier the fleet deployment plan of a specific fleet size and mix strategy n in period t is dependent on the container shipment demand of the previous period $t-1$. Therefore, given a fleet size and mix strategy n in period t, which is produced by a parent m in period $t-1$, the optimal fleet deployment plan for strategy n is dependent on the container shipment demand scenario s' in period $t-1$, which can be formulated as a 2SSP model with the objective of maximizing the expected profit across all container shipment demand scenarios s in period t, denoted by $EP_{t,n}^{m,s'}$.

It is noted that the decisions regarding δ_{nt}^{kr}, x_{nt}^{kr}, and y_{nt}^{k} are made prior to a realization of the random container shipment demand. In reality the number of containers transported between an O-D port pair $(o, d) \in \mathcal{W}$ assigned to a particular container route can be determined only after the realization of the random container shipment demand. We can thus break down the set of all decision variables into two stages. The first-stage decision variables are δ_{nt}^{kr},

x_{nt}^{kr}, and y_{nt}^{k}, and the second-stage variables are $z_{snt}^{h^{od}}$. Therefore the 2SSP model is as follows:

[2SSP]

$$EP_{t,n}^{m,s'} = \max \sum_{k \in \mathcal{G}_{t,n}^{\mathrm{OUT}}} c_{kt}^{\mathrm{OUT}} + \sum_{k \in \mathcal{G}_{t,n}^{\mathrm{SOLD}}} c_{kt}^{\mathrm{SOLD}} - \sum_{r \in \mathcal{R}} \sum_{k \in \mathcal{G}_{t,n}} c_{krt} x_{nt}^{kr} - \sum_{k \in \mathcal{G}_{t,n}} e_{kt} y_{nt}^{k} - \sum_{k \in \mathcal{G}_{t,n}^{\mathrm{IN}}} c_{kt}^{\mathrm{IN}}$$
$$- \sum_{k \in \mathcal{G}_{t,n}^{\mathrm{NEW}}} c_{kt}^{\mathrm{NEW}} + \sum_{s \in \mathcal{S}_t} p_{s|s'}^{t} Q_{\boldsymbol{\xi}}^{ts}(\mathbf{v}, \boldsymbol{\xi}(\boldsymbol{\omega})) \tag{11.10}$$

subject to

$$\delta_{nt}^{kr} \le x_{nt}^{kr} \le M^{kr} \delta_{nt}^{kr}, \quad \forall r \in \mathcal{R}, \forall k \in \mathcal{G}_{t,n} \tag{11.11}$$

$$\sum_{k \in \mathcal{K}} x_{kr} \ge N_r, \quad \forall r \in \mathcal{R} \tag{11.12}$$

$$\Delta t - T_k^t \le y_{nt}^{k}, \quad \forall k \in \mathcal{G}_{t,n} \tag{11.13}$$

$$x_{nt}^{kr} t_t^{kr} + y_{nt}^{k} = \Delta t, \quad \forall r \in \mathcal{R}, \forall k \in \mathcal{G}_{t,n} \tag{11.14}$$

$$\sum_{r \in \mathcal{R}} \delta_{nt}^{kr} = 1, \quad \forall k \in \mathcal{G}_{t,n} \tag{11.15}$$

$$x_{nt}^{kr} \in \mathbb{Z}^{+} \cup \{0\}, \quad \forall k \in \mathcal{K}, \forall r \in \mathcal{R} \tag{11.16}$$

$$y_{nt}^{k} \ge 0, \quad \forall k \in \mathcal{K} \tag{11.17}$$

$$\delta_{nt}^{kr} = \{0, 1\}, \quad \forall k \in \mathcal{G}_{t,n}, \forall r \in \mathcal{R} \tag{11.18}$$

where, for succinctness, $\mathbf{v} = \left(\cdots \delta_{nt}^{kr} \cdots x_{nt}^{kr} \cdots y_{nt}^{k} \cdots\right)$ contains all first-stage decision variables; M^{kr} represents the maximum number of voyages ship k can complete on route r during period t; N_r is the number of voyages required on route r during period t in order to maintain a given level of service frequency; Δt is the duration of period t (days); T_k^t represents the shipping season for ship k in period t (days), referring to the number of days that it is safe and appropriate for the ship to sail on the sea; t^{kr} is the voyage time of ship k on route r (days/voyage); and $\mathbb{Z}^{+}$ is the set of positive integers. $Q_{\boldsymbol{\xi}}^{ts}(\mathbf{v}, \boldsymbol{\xi}(\boldsymbol{\omega}))$ is a function used for the following second-stage optimization problem, which depends on the first-stage decision variables and the realization of container shipment demand $\boldsymbol{\omega}$; under scenario s, its value is obtained by solving the following optimization problem:

$$Q_{\boldsymbol{\xi}}^{ts}(\mathbf{v}, \boldsymbol{\xi}(\boldsymbol{\omega})) = \max \sum_{(o,d) \in \mathcal{W}} \sum_{h^{od} \in \mathcal{H}^{od}} \left(f_t^{od} - c_t^{h^{od}}\right) z_{snt}^{h^{od}} \left(\omega_{st}^{od}\right) \tag{11.19}$$

subject to

$$\sum_{k\in\mathcal{G}_{t,n}} x_{nt}^{kr} V_k \geq \sum_{(o,d)\in\mathcal{W}} \sum_{h^{od}\in\mathcal{H}^{od}} \rho_{ir}^{h^{od}} z_{snt}^{h^{od}}\left(\omega_{st}^{od}\right), \quad \forall i=1,\ldots,m_r, \forall r\in\mathcal{R}, \forall s\in\mathcal{S}_t \tag{11.20}$$

$$\sum_{h^{(o,d)}\in\mathcal{H}^{(o,d)}} z_{snt}^{h^{od}}\left(\omega_{st}^{od}\right) \leq \xi^{od}\left(\omega_{st}^{od}\right), \quad \forall(o,d)\in\mathcal{W}, \forall s\in\mathcal{S}_t \tag{11.21}$$

$$z_{snt}^{h^{od}} \geq 0, \forall(o,d)\in\mathcal{W}, \quad \forall h^{od}\in\mathcal{H}^{od}, \forall s\in\mathcal{S}_t \tag{11.22}$$

where V_k is the capacity of a particular ship k (TEUs), $\rho_{ir}^{h^{od}}$ is a binary coefficient that equals 1 if a container route $h^{od}\in\mathcal{H}^{od}$ contains leg i of route r and otherwise equals 0.

Eq. (11.10) is the objective function of the 2SSP model. The set of constraints (11.20) applies the big-M method to ensure that if δ_{tn}^{kr} equals 0, then x_{tn}^{kr} equals 0; otherwise, if δ_{tn}^{kr} equals 1 then x_{tn}^{kr} would be a positive integer. The value of Mkr can be given by $M^{kr}=\lfloor\Delta t/t^{kr}\rfloor$, where $\lfloor a\rfloor$ denotes the maximum integer not greater than a. The set of constraints (11.12) gives the number of voyages required on route r in order to maintain a given level of liner shipping frequency. For example, if a weekly shipping service is required on each liner ship route during a planning horizon of six months, then $N_r=26$. The set of constraints (11.13) provide the minimum lay-up days of ship k on route r. Eq. (11.14) indicates that the total voyage time of ship k on route r at sea and the lay-up time should not exceed one single period. Eq. (11.15) ensures that each ship only serves on one route. Constraints (11.16)–(11.18) defines the range of decision variables, x_{nt}^{kr}, y_{nt}^{k}, and δ_{nt}^{kr}, respectively.

Eq. (11.19) is the objective function of the second-stage optimization problem. The left-hand side of the constraints (11.20) is the total transportation capacity of ships deployed on the liner ship route $r\in\mathcal{R}$. The right-hand side of the constraints (11.20) is the total number of containers carried by ships sailing on leg l of the liner ship route $r\in\mathcal{R}$, including the containers loaded at previously ports of call, which have remained on the ships plus any containers loaded or transshipped at port p_r^i. Therefore the constraints (11.20) guarantee the total number of containers transported on each leg of a liner shipping service route does not exceed the ship capacity deployed on the route. The constraints (11.21) imply that the total number of containers assigned to all the ship routes between an O-D port pair does not exceed the corresponding container shipment demand. Eq. (11.22) requires the decision variables $z_{snt}^{h^{od}}$ should be nonnegative.

After $EP_{t,n}^{m,s'}$ is obtained by solving the 2SSP model above, we can then calculate the expected profit under strategy n in period t, given strategy m was applied in period $t-1$, which is denoted by $EP_{t,n}^{m}$ $(t=1,\ldots,T; n=1,\ldots,N_t)$, and given by:

$$EP_{t,n}^{m} = \sum_{s'\in\mathcal{S}_{t-1}} p_{s'}^{t-1} \times EP_{t,n}^{m,s'} \tag{11.23}$$

Multiperiod Stochastic Programming Model

At the end of period T the set of ships owned by the liner container shipping company under strategy n°, denoted by $\bar{\mathcal{G}}_{T,n^\circ}$, includes ships that were kept, ships that were chartered out, and ships that were bought at the beginning of period T:

$$\bar{\mathcal{G}}_{T,n^\circ} = \mathcal{G}_{T,n^\circ}^{\text{KEEP}} \cup \mathcal{G}_{T,n^\circ}^{\text{OUT}} \cup \mathcal{G}_{T,n^\circ}^{\text{NEW}}, \quad n^\circ = 1,\ldots,N_T \tag{11.24}$$

For simplicity, we assume that all ships owned by the liner container shipping company are disposed of at the end of period T for their salvage values, which are denoted by SV_{T,n°. The objective of the MPLSFP problem is to find the best policy that maximizes the sum of the expected profits across the whole T-period planning horizon plus the salvage value. Here a policy refers to a path from the dummy root O to the leaf node $n^\circ \in \mathcal{N}_T = \{1, \ldots, N_T\}$ in the decision tree. Therefore the best policy refers to the path from the dummy root O to a leaf node n° in the decision tree, with the maximal sum of expected profits plus salvage values. The length of a path is, as usual, the sum of the length of the arcs that it contains.

Let $\mathcal{L}_{t,n}^{n^\circ,l}$ be 1 if a path $l \in \mathbb{P}^T(n^\circ)$ from the dummy root O to the leaf node n° passes node n of period t; otherwise, it is 0 $(n^\circ = 1,\ldots,NT)$. The best path with the maximal sum of expected profits across all periods plus salvage value (i.e., the optimal plan for the MPLSFP problem) is given by:

$$Z = \max_{\substack{l \in \mathbb{P}^T(n^\circ) \\ n^\circ \in \mathcal{N}_T}} \sum_{n,m\in\mathbb{T}^0} \sum_{t=1}^{T} \frac{EP_{t,n}^{m} \mathcal{L}_{t,n}^{n^\circ,l}}{(1+r)^t} + \frac{SV_{T,n^\circ}}{(1+r)^T} \tag{11.25}$$

where r is the discount rate of each period during the multiperiod planning horizon.

SOLUTION ALGORITHM

As shown in Fig. 11.3, the expected profit on each arc contributes to the total profits along a given path from the dummy root O to a leaf node n°. In order to find the path with the greatest total profits across all periods, the attribute of each arc $EP_{t,n}^m$ and the salvage value SV_{T,n° have to be obtained. Once each $EP_{t,n}^m$ is obtained, the path from the dummy root O to a leaf node n° with the maximal total profit can be found. Therefore the key aspect of the solution method is to obtain $EP_{t,n}^m$, which is to solve the 2SSP model. The following firstly proposes a solution method to deal with the 2SSP model in order to get $EP_{t,n}^m$, then describes an algorithm for finding the best path for the proposed MPLSFP problem in this chapter.

Dual Decomposition and Lagrangian Relaxation Method for Solving 2SSP Models

It is noted that each 2SSP model under strategy n for period t involves a number of scenarios of the uncertain container shipment demand. Even when the first-stage decisions are given and fixed, S_t $(t=1,\ldots,T)$ optimization models (11.19) have to be solved in order to obtain the expected value associated with this given set of fixed first-stage decisions.

In order to effectively solve a 2SSP model under strategy n for period t $(n=1,\ldots,Nt;\ t=1,\ldots,T)$, the dual decomposition and Lagrangian relaxation method proposed by Carøe and Schultz (1999) is used because it can decompose the 2SSP model into S_t sub-problems based on the scenarios of container shipment demand. In order to do that, the first-stage variables are copied for each scenario. Such duplication might result in a new problem: The first-stage decision variables vs for each scenario s $(s=1,\ldots,St)$ could be different. However, the first-stage decision variable vector vs $(s=1,\ldots,St)$ in the 2SSP model should be independent of uncertain container shipment demand because they are made prior to knowing the exact market demand. Therefore the nonanticipativity constraints $\mathbf{v}_1=\mathbf{v}_2=\cdots=\mathbf{v}_{S_t}(t=1,\ldots,T)$ are added so as to guarantee that the first-stage decisions in period t do not depend on the scenarios. The nonanticipativity constraints are implemented through the equation $\sum_{s\in\mathcal{S}_t}\mathbf{H}^s\mathbf{v}^s=\mathbf{0}$ $(t=1,\ldots,T)$, where $\mathbf{H}^s$ is a suitable matrix with $(S_t-1)\times(2K_{tn}R+K_{tn})$ rows and $2K_{tn}R+K_{tn}$ columns (K_{tn} is the cardinality of set $\mathcal{G}_{t,n}$, namely the number of ships; $2K_{tn}R+K_{tn}$ is the number of first-stage decision variables x_{nt}^{kr}, y_{nt}^{k} and δ_{nt}^{kr}) for $s = 1,\ldots,St$, defined as follows:

$$\mathbf{H}^1 = (\mathbf{I}, \mathbf{0}, \ldots, \mathbf{0})', \ \mathbf{H}^2 = (-\mathbf{I}, \mathbf{I}, \mathbf{0}, \ldots \mathbf{0})', \ \mathbf{H}^3 = (\mathbf{0}, -\mathbf{I}, \mathbf{I}, \ldots \mathbf{0})', \ldots,$$
$$\mathbf{H}^{S_t - 1} = (\mathbf{0}, \ldots, -\mathbf{I}, \mathbf{I})', \ \mathbf{H}^{S_t} = (\mathbf{0}, \ldots \mathbf{0}, -\mathbf{I})' \tag{11.26}$$

where I and 0 are the square unity matrix and zero matrix of size $2K_{tn}R + K_{tn}$, respectively. Let $\boldsymbol{\lambda}$ be a $(S_t - 1) \times (2K_{tn}R + K_{tn})$ – dimensional vector of the Lagrangian multiplier associated with the nonanticipativity constraints. The resulting Lagrangian relaxation is as follows:

[LRt,n]

$$LR_{t,n}(\boldsymbol{\lambda}) = \max \sum_{s \in \mathcal{S}_t} p_{s|s'}^t \left(\begin{array}{l} \sum\limits_{k \in \mathcal{G}_{t,n}^{\mathrm{OUT}}} c_{kt}^{\mathrm{OUT}} + \sum\limits_{k \in \mathcal{G}_{t,n}^{\mathrm{SOLD}}} c_{kt}^{\mathrm{SOLD}} - \sum\limits_{k \in \mathcal{G}_{t,n}^{\mathrm{IN}}} c_{kt}^{\mathrm{IN}} - \sum\limits_{k \in \mathcal{G}_{t,n}^{\mathrm{NEW}}} c_{kt}^{\mathrm{NEW}} \\ - \sum\limits_{r \in \mathcal{R}} \sum\limits_{k \in \mathcal{G}_{t,n}} c_{krt} x_{nt}^{krs} - \sum\limits_{k \in \mathcal{G}_{t,n}} e_{kt} y_{nt}^{ks} + Q_{\boldsymbol{\xi}}^{ts}\left(\mathbf{v}^s, \boldsymbol{\xi}(\boldsymbol{\omega})\right) \end{array} \right) + \boldsymbol{\lambda}' \mathbf{H}^s \mathbf{v}^s \tag{11.27}$$

subject to constraints (11.11)–(11.18), (11.20)–(11.22) for each scenario of container shipment demand. This Lagrangian relaxation model LRt,n can be further decomposed into S_t separate mixed-integer linear programming problems according to the S_t container shipment demand scenarios, namely:

$$LR_{t,n}(\boldsymbol{\lambda}) = \sum_{s \in S_t} LR_{t,n}^s(\boldsymbol{\lambda}) \tag{11.28}$$

where

$$LR_{t,n}^s(\boldsymbol{\lambda}) = \max p_{s|s'}^t \left(\begin{array}{l} \sum\limits_{k \in \mathcal{G}_{t,n}^{\mathrm{OUT}}} c_{kt}^{\mathrm{OUT}} + \sum\limits_{k \in \mathcal{G}_{t,n}^{\mathrm{SOLD}}} c_{kt}^{\mathrm{SOLD}} - \sum\limits_{k \in \mathcal{G}_{t,n}^{\mathrm{IN}}} c_{kt}^{\mathrm{IN}} - \sum\limits_{k \in \mathcal{G}_{t,n}^{\mathrm{NEW}}} c_{kt}^{\mathrm{NEW}} \\ - \sum\limits_{r \in \mathcal{R}} \sum\limits_{k \in \mathcal{G}_{t,n}} c_{krt} x_{nt}^{krs} - \sum\limits_{k \in \mathcal{G}_{t,n}} e_{kt} y_{nt}^{ks} + Q_{\boldsymbol{\xi}}^{ts}(\mathbf{v}^s, \boldsymbol{\xi}(\boldsymbol{\omega})) \end{array} \right) + \boldsymbol{\lambda}' \mathbf{H}^s \mathbf{v}^s \tag{11.29}$$

subject to constraints (11.11)–(11.18), (11.20)–(11.22) associated with the sth scenario of container shipment demand.

Each subproblem shown in Eq. (11.29) can be solved efficiently using an optimization solver of linear integer programs such as CPLEX. It is straightforward to demonstrate that $LR_{t,n}(\boldsymbol{\lambda})$, the objective function value of the LRt,n model with respect to a given Lagrangian multiplier $\boldsymbol{\lambda}$, is an upper bound on the optimal value of Eq. (11.10). The best or tightest upper bound is found by solving the Lagrangian dual:

[LDt,n]

$$LD_{t,n} = \min_{\boldsymbol{\lambda}} LR_{t,n}(\boldsymbol{\lambda}) \tag{11.30}$$

which is solved by the subgradient method; that is, a brazen adaptation of the gradient method in which gradients are replaced by subgradients. Carøe and Schultz (1999) have shown that $\sum_{s \in \mathcal{S}_t} \mathbf{H}^s \mathbf{v}^{s^*}$ is the subgradient of (11.27), where $\mathbf{v}^{s^*}$ is the optimal solution of the sth subproblem (11.29). With this subgradient the LRt,n model can be solved using the following subgradient method:

Step 0: Give an initial Lagrangian multiplier vector $\boldsymbol{\lambda}^{(1)}$. Let the number of iterations $h = 1$.

Step 1: Calculate the subgradient $\sum_{s \in \mathcal{S}_t} \mathbf{H}^s \mathbf{v}^{s^*(h)}$ by solving the subproblem shown in Eq. (11.29) with respect to the Lagrangian multiplier vector $\boldsymbol{\lambda}^{(h)}$.

Step 2: Update the Lagrangian multiplier vector according to the formula:

$$\boldsymbol{\lambda}^{h+1} = \boldsymbol{\lambda}^h + \kappa^h \sum_{s \in \mathcal{S}_t} \mathbf{H}^s \mathbf{v}^{s^*(h)} \tag{11.31}$$

where κ^h is a positive scalar step size and is given by

$$\kappa^h = 1/h \tag{11.32}$$

Step 3: If the following criterion is fulfilled, the algorithm is terminated. Otherwise, let $h = h + 1$ and go to Step 1.

$$\left| \left(LR_{t,n}(\boldsymbol{\lambda}^{h+1}) - LR_{t,n}(\boldsymbol{\lambda}^h) \right) / LR_{t,n}(\boldsymbol{\lambda}^h) \right| \le \tau \tag{11.33}$$

Shortest Path Algorithm for the Multiperiod LSFP Problem

Once the attribute of each arc has been obtained using the solution method described above, the next step is to find the longest path from the dummy root O to a leaf node, using the maximal profit (summed across all arcs contained in this path) plus the salvage value. Each leaf node, n^o, is connected to a dummy destination node D (shown in Fig. 11.4) by a dummy arc, and the

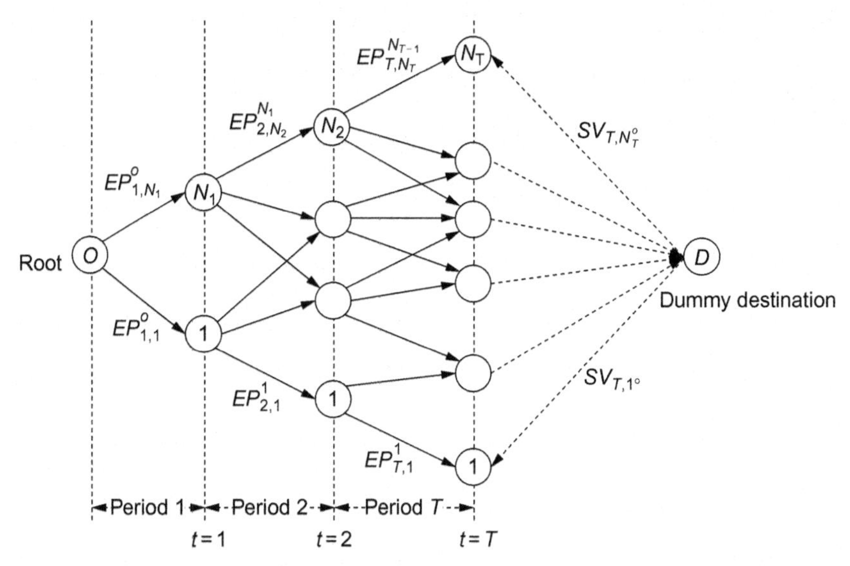

Fig. 11.4 An acyclic network representation.

value on each dummy arc is set equal to the salvage value of the leaf node, SV_{T,n°.

Next, finding the longest path from the dummy root O to a leaf node is equivalent to finding the longest path from O to D in the acyclic network, shown in Fig. 11.4. Any shortest-path algorithms applicable to an acyclic network can be applied to find the longest path, such as Dijkstra's algorithm (Ahuja, Magnanti, & Orlin, 1996). It is noted that in order to use shortest path algorithms, a negative sign is added to the attribute of each branch, that is we consider $-EP^{m}_{t,n}$. Then the shortest path, found using shortest-path algorithms, is actually the longest path that we are seeking.

COMPUTATIONAL RESULTS

Numerical Example Design

In order to illustrate the applicability of the proposed approach to the MPLSFP problem with container transshipment and demand uncertainty, we provide a numerical example, using the liner shipping topology and 36 calling ports depicted in Fig. 8.2. In the example, we assume that a liner container shipping company intends to make a 10-year liner ship fleet plan for providing a weekly liner shipping service.

Table 11.2 Ship Fleet at the Beginning of Research Horizon

	Ship Types				
Item	**1**	**2**	**3**	**4**	**5**
Ship size (TEUs)	2808	3218	4500	5714	8063
Design speed (knots)	21.0	22.0	24.2	24.6	25.2
Daily operating cost (10^3 \$)	19.8	22.5	30.9	38.8	54.2
Daily lay-up cost (10^3 \$)	2.8	3.2	4.5	6	8
Annual chartering out rate (million \$)	3.64	4.68	6.42	8.64	10.24
Annual chartering in rate (million \$)	4	5.2	7.0	9.4	12.0
Selling price (million \$)	85	105	175	225	345
Purchasing price (million \$)	135	155	215	275	385

The relevant ship data are presented in Table 11.2, including ship size and type, daily operating and lay-up costs, annual chartering in and out rates, and selling and purchasing prices. To simplify the input data preparations, it is assumed that these cost parameters do not change within the time horizon of this fictitious example.

Generation of Demand Scenarios and Fleet Size and Mix Strategies

We assume there are three scenarios of container shipment demand: high, medium, and low for each single period (i.e., one year) shown in Fig. 11.5. Additionally, we assume three feasible strategies are proposed by the liner container shipping company's experts at the beginning of each year (see Table 11.3). A strategy involves five options: keep, charter out, sell, charter in, and buy ships. We use five capital letters, *K, O, S, I,* and *B,* to represent those five options, respectively. Additionally the superscript and

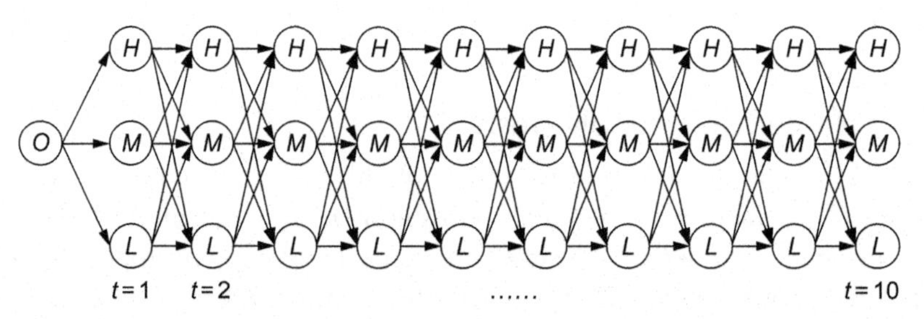

Fig. 11.5 Scenario tree for the numerical example.

Table 11.3 Strategies Proposed for Each Year

Year	Scenario 1	Scenario 2	Scenario 3
1	$K_2^1K_2^2K_9^3K_2^4K_{12}^5$	$K_2^1K_2^2K_9^3I_1^3K_2^4K_{12}^5$	$K_1^1O_1^1K_2^2K_9^3I_4^3K_2^4K_{12}^5$
2	$K_2^1K_2^2K_9^3I_5^3K_2^4K_{12}^5$	$K_2^1K_2^2K_9^3I_2^3K_2^4I_2^4K_{12}^5$	$K_2^1K_2^2K_9^3I_3^3K_2^4I_2^4K_{12}^5$
3	$S_1^1K_1^1K_2^2K_9^3I_4^3K_2^4I_2^4K_{12}^5$	$S_2^1K_2^2K_9^3B_5^3K_2^4K_{12}^5$	$S_1^1K_1^1K_2^2K_9^3K_2^4I_6^4K_{12}^5$
4	$K_1^1K_2^2K_9^3I_4^3K_2^4I_5^4K_{12}^5$	$K_2^2K_{14}^3K_2^4K_{12}^5$	$K_1^1K_2^2K_9^3B_5^3K_2^4I_3^4K_{12}^5$
5	$S_1^1K_2^2K_9^3B_5^3K_2^4I_3^4K_{12}^5$	$K_2^2K_{14}^3K_2^4I_5^4K_{12}^5$	$K_1^1K_2^2K_{14}^3K_2^4I_5^4K_{12}^5$
6	$K_2^2K_{14}^3K_2^4I_8^4K_{12}^5$	$S_1^2K_1^2K_{14}^3K_2^4B_4^4K_{12}^5I_2^5$	$S_1^1K_2^2K_{14}^3K_2^4B_4^4K_{12}^5$
7	$S_2^2K_{14}^3K_2^4B_4^4K_{12}^5I_3^5$	$K_1^2K_{14}^3K_6^4K_{12}^5I_5^5$	$K_2^2K_{14}^3K_6^4K_{12}^5I_5^5$
8	$K_{14}^3K_6^4K_{12}^5I_6^5$	$S_1^2K_{14}^3K_6^4K_{12}^5B_6^5$	$S_1^2K_1^2K_{14}^3K_6^4K_{12}^5B_6^5$
9	$K_{14}^3K_6^4I_4^4K_{12}^5B_3^5$	$K_{14}^3K_6^4K_{18}^5$	$S_1^2K_{14}^3K_6^4K_{18}^5$
10	$K_{14}^3K_6^4K_{15}^5I_5^5$	$K_{14}^3K_6^4I_3^4K_{18}^5$	$K_{14}^3K_6^4I_4^4K_{18}^5$

the subscript of the capital letters in a strategy represent the ship type and the number of ships of this type, respectively. For example, the strategy $K_2^1K_2^2K_9^3I_1^3K_2^4K_{12}^5$ in year 1 indicates that a total of 28 ship are contained in the ship fleet, of which two ships of type 1, two ships of type 2, nine ships of type 3, two ships of type 4, and 12 ships of type 5 are kept in the ship fleet; while one ship of type 3 is chartered in.

Profit Comparison

The results of the numerical example are illustrated as an acyclic network representation. It is found that the longest path from O to D is $O \rightarrow 1 \rightarrow 3 \rightarrow 3 \rightarrow 1 \rightarrow 1 \rightarrow 2 \rightarrow 2 \rightarrow 2 \rightarrow 2 \rightarrow 3 \rightarrow D$, with total profits of 95.2586 billion dollars.

As already mentioned the most significant contribution of this study is to take the dependency of uncertain container shipment demand between periods into account in the multiperiod LSFP problem. In order to evaluate whether it is worthwhile when considering container shipment demand dependency, as well as to investigate the effect of the dependency on profit, we compute the total profit over the whole multiperiod planning horizon for the same numerical example. We assume that the container shipment demand in each period is independent of that in other periods and compare the results with those produced above. For the sake of presentation, in the remainder of this paper, the case regarding dependency of container shipment demand is called case I (i.e., the problem studied in this paper), while the case with independent container shipment demand is called case II.

In case II, $EP_{t,n}^{m}$ $(t=1,\ldots,T; n=1,\ldots,N_t)$ is given by:

$$EP_{t,n}^{m} = \max \sum_{k\in\mathcal{G}_{t,n}^{\text{OUT}}} c_{kt}^{\text{OUT}} + \sum_{k\in\mathcal{G}_{t,n}^{\text{SOLD}}} c_{kt}^{\text{SOLD}} - \sum_{r\in\mathcal{R}}\sum_{k\in\mathcal{G}_{t,n}} c_{krt}x_{nt}^{kr} - \sum_{k\in\mathcal{G}_{t,n}} e_{kt}y_{nt}^{k}$$

$$- \sum_{k\in\mathcal{G}_{t,n}^{\text{IN}}} c_{kt}^{\text{IN}} - \sum_{k\in\mathcal{G}_{t,n}^{\text{NEW}}} c_{kt}^{\text{NEW}} + \sum_{s\in\mathcal{S}_t} p_s^{t} Q_{\xi}^{ts}(\mathbf{v},\xi(\boldsymbol{\omega})) \tag{11.34}$$

subject to constraints (11.11)–(11.22).

We found that the longest path from O to D for the numerical example was the same as the path in case I, but the total profit was 95.0217 billion dollars. The results show that the total profits in case II are lower than those in case I, which indicates that the dependency of container shipment demand has a significant effect on profits, verifying the importance of considering dependency between the container shipment demand in different periods. Actually, we have also theoretically proven that the profit in case II will be less than or equal to that in case I (see the following proposition).

Proposition 11.1. The profit using Eq. (11.23) in case I is larger or equal to that using Eq. (11.24) in case II.

Proof. In case I, $EP_{t,n}^{m}$ is given by:

$$EP_{t,n}^{m} = \sum_{s'\in\mathcal{S}_{t-1}} p_{s'}^{t-1} \times EP_{t,n}^{m,s'} \tag{11.35}$$

In Eq. (11.10), the terms $\sum_{k\in\mathcal{G}_{t,n}^{\text{OUT}}} c_{kt}^{\text{OUT}}$, $\sum_{k\in\mathcal{G}_{t,n}^{\text{SOLD}}} c_{kt}^{\text{SOLD}}$, $\sum_{k\in\mathcal{G}_{t,n}^{\text{IN}}} c_{kt}^{\text{IN}}$, and $\sum_{k\in\mathcal{G}_{t,n}^{\text{NEW}}} c_{kt}^{\text{BUY}}$ can be removed, as they are fixed when the sets of $\mathcal{G}_{t,n}^{\text{OUT}}, \mathcal{G}_{t,n}^{\text{SOLD}}, \mathcal{G}_{t,n}^{\text{IN}}$, and $\mathcal{G}_{t,n}^{\text{NEW}}$ are given. Then Eq. (11.35) could be rewritten as follows, after substituting Eq. (11.10) to replace $EP_{t,n}^{m,s'}$:

$$
\begin{aligned}
EP_{t,n}^{m} &= \sum_{s'\in\mathcal{S}_{t-1}} p_{s'}^{t-1} \times \max\left(\sum_{s\in\mathcal{S}_t} p_{s|s'}^{t} Q_{\xi}^{ts}(\mathbf{v},\boldsymbol{\xi}(\boldsymbol{\omega})) - \sum_{r\in\mathcal{R}}\sum_{k\in\mathcal{G}_{t,n}} \left(c_{kt}^{r}x_{nt}^{kr} + e_{kt}y_{nt}^{k}\right)\right)\\
&= \sum_{s'\in\mathcal{S}_{t-1}} \max p_{s'}^{t-1} \times \left(\sum_{s\in\mathcal{S}_t} p_{s|s'}^{t} Q_{\xi}^{ts}(\mathbf{v},\boldsymbol{\xi}(\boldsymbol{\omega})) - \sum_{r\in\mathcal{R}}\sum_{k\in\mathcal{G}_{t,n}} \left(c_{kt}^{r}x_{nt}^{kr} + e_{kt}y_{nt}^{k}\right)\right)\\
&\geq \max \sum_{s'\in\mathcal{S}_{t-1}} p_{s'}^{t-1} \times \sum_{s\in\mathcal{S}_t} p_{s|s'}^{t} Q_{\xi}^{ts}(\mathbf{v},\boldsymbol{\xi}(\boldsymbol{\omega})) - \sum_{r\in\mathcal{R}}\sum_{k\in\mathcal{G}_{t,n}} \left(c_{kt}^{r}x_{nt}^{kr} + e_{kt}y_{nt}^{k}\right)\\
&= \max \sum_{s\in\mathcal{S}_t}\sum_{s'\in\mathcal{S}_{t-1}} p_{s'}^{t-1} \times p_{s|s'}^{t} Q_{\xi}^{ts}(\mathbf{v},\boldsymbol{\xi}(\boldsymbol{\omega})) - \sum_{r\in\mathcal{R}}\sum_{k\in\mathcal{G}_{t,n}} \left(c_{kt}^{r}x_{nt}^{kr} + e_{kt}y_{nt}^{k}\right)\\
&= \max \sum_{s\in\mathcal{S}_t} p_{s}^{t} \times Q_{\xi}^{ts}(\mathbf{v},\boldsymbol{\xi}(\boldsymbol{\omega})) - \sum_{r\in\mathcal{R}}\sum_{k\in\mathcal{G}_{t,n}} \left(c_{kt}^{r}x_{nt}^{kr} + e_{kt}y_{nt}^{k}\right)
\end{aligned}
\tag{11.36}
$$

In case II, $EP_{t,n}^{m}$ is given by Eq. (11.34). Similarly, the terms $\mathcal{G}_{t,n}^{\mathrm{OUT}}, \mathcal{G}_{t,n}^{\mathrm{SOLD}}, \mathcal{G}_{t,n}^{\mathrm{IN}}$, and $\mathcal{G}_{t,n}^{\mathrm{NEW}}$ are removed and then $EP_{t,n}^{m}$ is given by:

$$
EP_{t,n}^{m} = \max \sum_{s\in\mathcal{S}_t} p_{s}^{t} Q_{\xi}^{ts}(\mathbf{v},\boldsymbol{\xi}(\boldsymbol{\omega})) - \sum_{r\in\mathcal{R}}\sum_{k\in\mathcal{G}_{t,n}} \left(c_{kt}^{r}x_{nt}^{kr} + e_{kt}y_{nt}^{k}\right) \tag{11.37}
$$

Therefore $EP_{t,n}^{m}$ in case I $\geq EP_{t,n}^{m}$ in case II.□

Comparison of Fleet Deployment Plans

The above section evaluates whether it is worthwhile to consider shipment demand dependency by investigating the effect of this dependency on profit. Similarly, this section investigates the effect of the dependency on the resulting fleet deployment plans. The 2SSP model (11.10) indicates that the fleet deployment plan under a given fleet strategy n in period t is dependent on the container shipment demand scenario s' of the previous period $t-1$. As there are S_{t-1} container shipment demand scenarios in period $t-1$, it is possible that there are S_{t-1} different fleet deployment plans for a strategy n in year t ($t=2,\ldots,T$), where each fleet deployment plan corresponds to a container shipment demand scenario s' from the previous period $t-1$ and is obtained by solving the 2SSP model (11.10). This shows that in case I the fleet deployment decisions for period t take the container shipment demand from the previous period into account, therefore the fleet deployment plans are demand dependent. In case II, the container shipment demand between periods is assumed to be independent; that is, the container shipment demand in period $t-1$ is not taken into consideration in the fleet deployment plan developed for period t, which indicates that the fleet deployment plans are demand independent. The optimization model (11.34) shows that in case II a strategy n in year t ($t=2,\ldots,T$) has only one fleet deployment plan, which is obtained by solving the optimization model. Obviously the demand-dependent fleet deployment plans in case I are more reasonable and flexible, because the consideration of container shipment demand dependency means that the liner container shipping company can adopt a proper fleet deployment plan based on the container shipment demand that came about in the previous period. Meanwhile, in case II the same fleet deployment plan must be adopted regardless of the scenario of container shipment demand that materialized the previous year.

In the numerical example, each fleet strategy has three fleet deployment plans corresponding to three scenarios of demand: high, medium, and low. For example, for the strategy $K_2^1K_2^2K_9^3I_5^3K_2^4K_{12}^5$ of year 2 in case I, three fleet deployment plans are shown in Table 11.4. The fleet deployment plan for the same strategy in case II is shown in Table 11.5. It is found that those fleet deployment plans are different; the reason for this is that the probabilities involved in the optimization models are different.

Table 11.4 Ship-to-Route Allocation of Strategy $K_2^1K_2^2K_9^3I_5^3K_2^4K_{12}^5$ in Case I for Year 2

		Route Type							
Demand Scenario	**Ship Type**	**CCX**	**CPX**	**GIS**	**IDX**	**NCE**	**NZX**	**SCE**	**UKX**
High	1							2	
	2							2	
	3	3	3			5	3		
	4			1	1				
	5			1	3	4		3	1
Medium	1							2	
	2							2	
	3	3	3			5	3		
	4					2			
	5			2	4	2		3	1
Low	1							2	
	2			2					
	3	3	3				4	4	
	4					1		1	
	5				4	6		1	1

Table 11.5 Ship-to-Route //cAllocation of Strategy $K_2^1K_2^2K_9^3I_5^3K_2^4K_{12}^5$ in Case II for Year 2

	Route Type							
Ship Type	**CCX**	**CPX**	**GIS**	**IDX**	**NCE**	**NZX**	**SCE**	**UKX**
1							2	
2							2	
3	3	3		2	2	3	1	
4					2			
5				3	4		2	1

SUMMARY

This chapter considers a multiperiod LSFP problem with container transshipment and uncertain container shipment demand. The uncertain container shipment demand in each period is assumed to be dependent on that of the previous period. A set of scenarios in each single period is used to reflect the uncertainty of container shipment demand; then the evolution and dependency of container shipment demand across multiple periods is modeled as a scenario tree. A decision tree is used to interpret the procedure of fleet development over the multiperiod planning horizon. Then the proposed multiperiod LSFP problem is formulated as a multiperiod stochastic programming model that comprises a sequence of interrelated 2SSP models. In order to solve this model the dual decomposition and Lagrangian relaxation method is employed to solve the 2SSP models; the solution to the multiperiod LSFP problem is then found by using a shortest path algorithm. We illustrate the applicability and performance of our proposed model and solution method on a numerical example. We also investigated the effect of container shipment demand dependency. The results show that the profit obtained when considering this dependency is higher; therefore the ship fleet plans are more flexible than when dependency is not considered.

It is worth highlighting the most significant contribution of this study, which is that it takes the first step towards a more realistic multiperiod LSFP problem than has been studied in previous literature and provides an applicable and feasible method for handling such a problem in practice. It has to be pointed out that rather than being regarded as decision variables, the feasible fleet size and mix strategies in each single period are assumed to be proposed by experts at the liner container shipping company. The rationale behind such an assumption is that it effectively reduces the searching space from the viewpoint of operation research and makes the multiperiod LSFP problem solvable in practice; otherwise, the multiperiod LSFP problem would be highly intractable. However, the quality of the solution (i.e., the longest path provided by this study) is relatively better than the others, but it is possibly not a global optimum. We also need to reduce the runtime further because the convergent rate of the harmonic series (i.e., the step size sequence adopted in the solution algorithm) is inefficient. It might be worthwhile to investigate whether a more sophisticated heuristic for finding feasible solutions would produce even better results.

REFERENCES

Ahmed, S., King, A. J., & Parija, G. (2003). A multi-stage stochastic integer programming approach for capacity expansion under uncertainty. *Journal of Global Optimization, 26*, 3–24.

Ahmed, S., & Sahinidis, N. V. (2003). An approximation scheme for stochastic integer programs arising in capacity expansion. *Operations Research, 51*(3), 461–471.

Ahuja, R. A., Magnanti, T. L., & Orlin, J. B. (1996). *Network flows: Theory, algorithms and applications*. Upper Saddle River, NJ: Prentice Hall.

Berman, O., Ganz, Z., & Wagner, J. M. (1994). A stochastic optimization model for planning expansion in a service industry under uncertain demand. *Naval Research Logistics, 41*, 545–564.

Carøe, C. C., & Schultz, R. (1999). Dual decomposition in stochastic integer programming. *Operations Research Letter, 24*, 37–45.

Celikyurt, U., & Özekici, S. (2007). Multiperiod portfolio optimization models in stochastic markets using the mean-variance approach. *European Journal of Operational Research, 179*, 186–202.

Cho, S. C., & Perakis, A. N. (1996). Optimal liner fleet routing strategies. *Maritime Policy and Management, 23*(3), 249–259.

Gülpinar, N., & Rustem, B. (2007). Worst-case robust decisions for multi-period mean-variance portfolio optimization. *European Journal of Operational Research, 183*, 981–1000.

Laguna, M. (1998). Applying robust optimization to capacity expansion of one location in telecommunications with demand uncertainty. *Management Science, 44*(11), 101–110.

Leung, S. C. H., Lai, K. K., Ng, W. -L., & Wu, Y. (2007). A robust optimization model for production planning of perishable products. *Journal of Operational Research Society, 58*(4), 413–422.

Leung, S. C. H., Tsang, S. O. S., Ng, W. -L., & Wu, Y. (2007). A robust optimization model for multi-site production planning problem in an uncertain environment. *European Journal of Operational Research, 181*(1), 224–238.

Listes, O., & Dekker, R. (2005). A scenario aggregation-based approach for determining a robust airline fleet composition for dynamic capacity allocation. *Transportation Science, 39* (3), 367–382.

Meng, Q., & Wang, T. (2011). A scenario-based dynamic programming model for multi-period liner ship fleet planning. *Transportation Research Part E-Logistics and Transportation Review, 47*(4), 401–413.

Nicholson, T. A. J., & Pullen, R. D. (1971). Dynamic programming applied to ship fleet management. *Operational Research Quarterly, 22*(3), 211–220.

Osorio, M. A., Gülpinar, N., & Rustem, B. (2008). A mixed integer programming model for multistage mean-variance post-tax optimization. *European Journal of Operational Research, 185*, 451–480.

Shapiro, A., & Philpott, A. (2007). *A tutorial on stochastic programming*. http://www2.isye.gatech.edu/people/faculty/Alex_Shapiro/TutorialSP.pdf.

Singh, K. J., Philpott, A. B., & Wood, P. K. (2009). Dantzig-Wolfe decomposition for solving multistage stochastic capacity-planning problems. *Operations Research*, *57*(5), 1271–1286.

Wagner, J. M., & Berman, O. (1995). Models for planning capacity expansion of convenience stores under uncertain demand and the value of information. *Annals of Operations Research*, *59*, 19–44.

Xie, X. L., Wang, T. F., & Chen, D. S. (2000). A dynamic model and algorithm for fleet planning. *Maritime Policy and Management*, *27*(1), 53–63.

PART FIVE

Conclusion

CHAPTER TWELVE

Conclusions and Future Outlook

Contents

CONCLUSIONS

This book addressed the need to investigate LSFP problems with container shipment demand uncertainty. A review of the current literature showed that there are many limitations and gaps in the LSFP problems studied so far. For example, there are no systematic methodologies proposed in the existing literature to deal with the uncertainty issue arising in LSFP problems. Besides this, the multiperiod LSFP problem is not properly addressed in the existing studies. This book worked on eliminating these limitations and gaps by proposing new methodologies. Also, solution algorithms were proposed in order to efficiently solve the new optimization models.

Part I introduced the shipping services and presented the liner ship fleet planning problems. Chapter 1 gave a comprehensive introduction to shipping services. Through this chapter, readers learned about the development of seaborne trade and its structure, the container trade, the maritime transportation and its three modes, and the shipping service and design of liner operations. Chapter 2 presented a critical literature review, focusing on three problems: fleet size and mix problems, fleet deployment problems, and fleet planning problems. Through this review, several potential problems and gaps were identified. Finally the chapter described the research topic of this book, namely that the uncertainty of container shipment demand has to be taken into account in the LSFP problems.

Part II introduces the methodology to deal with mathematical modeling in an uncertain environment. Chapter 3 focuses on the concept of stochastic

Liner Ship Fleet Planning
http://dx.doi.org/10.1016/B978-0-12-811502-2.00012-0

programming and discusses two basic approaches to modeling optimization problems with uncertainty: one is an expected value model and another is chance-constrained model. Based on the introduction of stochastic programming stated in Chapters 3, 4, and 5 presented the methodology to build a chance-constrained programming model and a two-stage stochastic programming model in details, respectively.

Part III introduced the solution algorithm to solve the chance-constrained programming model and a two-stage stochastic programming model presented in Part II. For chance-constrained programming models, Chapter 6 gave a comprehensive statement of the solution algorithm, which uses the average of the sample to approximate the chance-constrained programming problems in order to obtain good candidate solutions for the resulting problems. The convergence properties of the resulting problem was discussed as well. Chapter 7 focused on the solution algorithm to solve the two-stage stochastic programming model introduced in Chapter 5. The solution algorithm is to combine the dual decomposition scheme and Lagrangian relaxation jointly. The properties of the solution algorithms, including the convergence rate, the bounds, the quality of the solutions, and the performance of the algorithms are stated in Part III.

Part IV mainly illustrated the application of the models and solution algorithms stated in Part II and Part III in the LSFP problems with demand uncertainty. Chapter 8 made an initial investigation of the short-term LSFP problem, in which the container shipment demand is uncertain. To deal with the uncertainty, we assumed that the container shipment demand between each port pair on a liner ship route follows a normal distribution with a given mean and variance. This assumption may lead to a problem: As the demands are uncertain, once the decisions in the short-term LSFP problem have been determined, the fleet of ships may be unable to meet the pickup and delivery requirements of its customers, even though the expected demand along the routes does not exceed the fleet capacity. This is not completely avoidable, and the liner container shipping company can only hope that there is a very low possibility of it occurring.

In order to reduce the possibility that the liner container shipping company cannot satisfy the customers' demand, we decided to insist that the decisions guarantee feasibility "as much as possible." Therefore we formulated the constraint that the liner container shipping company satisfies customer demand in a probabilistic form in this chapter, which is called a chance constraint. The level of service was proposed to represent the probability of satisfying the customers' requirements, and this was formulated as a chance constraint. The short-term LSFP problem with an uncertain container shipment demand was then formulated as a CCP model. The model is actually an integer linear programming model, which can be solved efficiently by the optimization solver, CPLEX. A numerical example was implemented to evaluate the applicability of the proposed model and analyze the results corresponding to the confidence parameters with different values. The results implied that the level of service has a significant impact on the optimal fleet size and deployment. It was also found that more ships are needed and more costs must be incurred in order to maintain a higher level of service.

Chapter 9 explored the work described in Chapter 8 because the model proposed in Chapter 8 is based on the assumption that the demand of all port pairs are independent and follow normal distribution without verification, as well as that all ships have to be emptied at the start of each sailing voyage, which is not consistent with practice. Therefore the proposed LSFD problem in Chapter 9 is an extension to the study in Chapter 8 and formulated as a JCCP model. It has to note that the methodology described in Chapter 8 is restricted by a special assumption without verification that the container shipment demand of all port pairs are independent normally distributed; this study is without such a restriction and examines the container demand uncertainty by enforcing a service level for the shipping network. The model proposed in Chapter 8 can be regarded as a special case for the model proposed in Chapter 9, which is more practical because it is for the whole network.

Chapter 10 studied the short-term LSFP problem with container transshipment and uncertain container shipment demand from another point of view. The goal of this chapter was to maximize the expected profit for the liner container shipping company. The container shipment demand for each port pair on a liner ship route was assumed to be a random variable. This problem was reformulated as a 2SSIP model. In a stochastic model, some decisions have to be made before the uncertain terms are observed; these are known as first-stage decisions. Furthermore, after the uncertain terms become known, recourse actions can take place, which are called second-stage decisions. As the decisions about fleet design and deployment are made before considering container shipment demand, they should have first-stage decision variables, while the number of containers shipped between a port pair on a liner ship route should have second-stage decision variables. The objective of this 2SSIP model was to choose the first-stage decision variables, such as the numbers, types, and deployment of ships, in such a way that the sum of the first-stage profit and the expected value of the second-stage profit from shipping containers were maximized. To effectively solve the proposed model, the SAA method was first used to approximate the expected recourse function, then the dual decomposition and Lagrangian relaxation methods were used to solve the model. The applicability of the proposed model and the performance of solution algorithms in terms of solution quality and computational time were tested using a numerical example. The results indicate that the solution methods are effective. It was also found that the solution varies with the variability of the uncertain parameters. As the variability increases, the profit obtained by a liner container shipping company decreases.

Chapter 11 studied the long-term/multiperiod LSFP problem with container shipment demand uncertainty. This chapter proposed a more realistic problem for a liner container shipping company by taking the uncertainty and the dependency of container shipment demand into account. By using a scenario tree approach to model the evolution of dependent uncertain demand across two successive single periods, and also by using a decision tree to model the procedure of LSFP, the proposed problem was formulated as a multiperiod stochastic programming model, comprising a sequence of interrelated two-stage stochastic programming models developed for each single period. The two-stage stochastic model was solved by a method that combines the dual decomposition and Lagrangian relaxation. Each path from the root to a leaf on the decision tree corresponds to a multiperiod liner ship fleet plan. In order to evaluate the performance of the solution algorithm and whether the proposed model can be applied in practice, a numerical example was implemented. We compared the profit in Case I, in which dependency and uncertainty of container shipment demand were both included, to the profit in Case II, where only uncertainty was included. The results showed that the profit in Case I was higher than in Case II, indicating that case I was superior.

In short, the contributions of the book are as follows:

1. It proposes more realistic LSFP problems than have been studied previously in the literature.
2. It provides a fresh and worthwhile research area for classical LSFP problems by taking the uncertainty and dependence of container shipment demand into account in such problems.
3. It improves the existing mathematical programming models proposed for classical LSFP problems with deterministic container shipment demand.

4. It proposes algorithms and systematic methodologies for formulating LSFP problems with uncertain container shipment demand.
5. It provides an applicable and feasible way for a liner container shipping company to produce its liner ship fleet plans in practice.

Though the work in this book took the first step to consider the issue of demand uncertainty in the LSFP problem, the methodology of modeling and solution algorithm still can be improved. Recently, some researchers improved our work and proposed new and novel modeling for the LSFP problem with uncertain demand (see Ng, 2014, 2015). The new distribution-free optimization models proposed by Ng (2014, 2015) only require the specification of the mean, standard deviation, and an upper bound on the shipping demand and propose a new distribution-free optimization model.

OUTLOOK

This book provides many potential future research topics. Firstly, in this book, all containers are assumed to be of the same size, which indicates that they are all standard twenty-foot equivalent units (TEUs). In practice, however, there are multiple types of containers with different sizes and weights, such as eight-foot equivalent units (EEUs), forty-foot equivalent units (FEUs), refrigerated containers, high cube containers, flat rack containers, platform containers, and others. The combination of these multiple types of container makes the operation of loading and unloading them much more complicated, and their inclusion in the LSFP problem would make it more realistic. This would be an interesting and worthwhile research area.

Secondly, although the solution algorithm that integrates the sample average approximation approach with a dual decomposition technique, as proposed in Chapter 10, can produce high-quality results, its convergence speed in approaching the optimum is slow; thus the computational time required is unsatisfactory. How one could increase the convergence speed and reduce the computational time would be a further interesting and challenging issue to explore. Distributed computing would be a useful tool for efficiently reducing the computational time (MirHassani et al., 2000).

Thirdly, Chapters 8, 9, 10, and 11 assume that the liner ship route network is predetermined and fixed. Such an assumption is reasonable in Chapters 8, 9, and 10 because of the short-term planning horizons of the LSFP problems studied in these chapters. During a short-term horizon a liner container shipping company would be unlikely (and probably unable) to change its liner ship route network. For the long-term LSFP problem, however, as studied in Chapter 11, the liner ship route network may not be fixed, as the company may change it. This assumption was made here in order to simplify the problem. In the future, however, we could extend this research work by integrating the dynamic routing problem with the long-term LSFP problem.

Fifthly, most liner trade is unbalanced because of the different economic needs in different regions. The number of inbound loaded containers can be quite different from the number of outbound loaded containers at any given port. Liner container shipping companies often need to reposition their empty containers or lease containers from vendors to meet customer demand. It would be interesting and worthwhile to look into repositioning empty containers and to discuss where and when companies should lease containers from vendors so as to meet the demand at different ports.

Finally, since ships are assets with finite lives, the liner container shipping company often has to consider which ships should be replaced and when. Therefore building a control model to capture ship utilization and replacement decisions would make the problem more realistic. The control model should jointly involve investment timing and trading strategies.

REFERENCES

MirHassani, S. A., Lucas, C., Mitra, G., Messina, E., & Poojari, C. A. (2000). Computational solution of capacity planning models under uncertainty. *Parallel Computing*, *26*, 511–538.

Ng, M. W. (2014). Distribution-free vessel deployment for liner shipping. *European Journal of Operational Research*, *238*(3), 858–862.

Ng, M. W. (2015). Container vessel fleet deployment for liner shipping with stochastic dependencies in shipping demand. *Transportation Research Part B*, *74*, 79–87.

INDEX

Note: Page numbers followed by *f* indicate figures, and *t* indicate tables.

E

F

H

I

J

L

M

N

O

P

Q

R

S

T

U

V

W

Printed in the United States
By Bookmasters